7

JAMES P. JOHNSON

The Politics

THE BITUMINOUS INDUSTRY FROM
WORLD WAR I THROUGH THE NEW DEAL

of Soft Coal

UNIVERSITY OF ILLINOIS PRESS
Urbana   Chicago   London

Library of Congress Cataloging in Publication Data

Johnson, James P.   1937–
    The politics of soft coal.

    Bibliography: p.
    Includes index.
    1. Coal trade—United States—History. 2. Bitu-
minous coal. 3. United States—Economic conditions—
To 1941. I. Title.
HD9545.J6   338.2′7′240973   78-31555
ISBN 0-252-00739-5

For Carolyn

Contents

Acknowledgments ix

Footnote Abbreviations for Manuscript Collections xi

1 Perspective 1

2 From Cooperation to Control 17

3 Stabilization under the Fuel Administration 53

4 From Riches to Rags 95

5 Drafting a Charter for Industrial Self-Government 135

6 The UMW and the New Deal 165

7 Price Stabilization under the Blue Eagle 195

8 The Failure of the Guffey Acts 217

9 The Politics of Soft Coal 239

Bibliographical Essay 247

Index 253

Acknowledgments

I WOULD LIKE to thank the many archivists and librarians who facilitated my research, particularly Joseph Howerton and Lane Moore at the National Archives, Robert Wood at the Herbert Hoover Presidential Library, and the staff at the Franklin D. Roosevelt Library. I owe sincere thanks to the Danforth Foundation, the Research Foundation of the City University of New York, and the National Coal Association for their assistance. Mrs. Nan C. Fahy superbly typed more drafts of this manuscript than either of us cares to recall.

I thank as well the *Journal of American History, The Register of the Kentucky Historical Society, Prologue: the Journal of the National Archives, Smithsonian,* Howard University Press and School of Social Science, Brooklyn College, City University of New York, for permission to use material in this book which they originally published in altered form as articles or papers.

Professors William E. Leuchtenburg, Robert D. Cross, Ellis W. Hawley, Ari Hoogenboom, Robert F. Himmelberg, Melvin I. Urofsky, and Robert Muccigrosso read and offered helpful criticism on sections of the manuscript. I want particularly to thank Professors Leuchtenburg and Hoogenboom for their stylistic suggestions and for their encouragement during the long gestation of this book. Professors Robert D. Cuff, Gerald D. Nash, and Jerome Sternstein read the entire manuscript and made observations which substantially influenced the result. All of these colleagues have convinced me that there really is a "community of scholars."

For reasons which they alone know, Stephen G. Young, Eugene L. Goldberg, my wife, Carolyn, and my daughters, Deborah and Katherine, all have my deepest gratitude.

Footnote Abbreviations
for Manuscript Collections

Baker Papers   Newton D. Baker Papers, Manuscript Division, Library of Congress

Brophy Papers   John Brophy Papers, Catholic University, Washington, D.C.

Daniels Papers   Josephus Daniels Papers, Manuscript Division, Library of Congress

Delano Papers   Frederick Delano Papers, Franklin D. Roosevelt Library, Hyde Park, New York

Garfield Papers   Harry A. Garfield Papers, Manuscript Division, Library of Congress

Hoover Papers   Herbert Hoover Papers, Herbert Hoover Presidential Library, West Branch, Iowa

Lauck Papers   William Jett Lauck Papers, University of Virginia Library, Charlottesville, Virginia

McAdoo Papers   William Gibbs McAdoo Papers, Manuscript Division, Library of Congress

Pinchot Papers   Gifford Pinchot Papers, Manuscript Division, Library of Congress

RG 9, NA   Record Group 9, "Records of the National Recovery Administration," National Archives

RG 60, NA   Record Group 60, "Department of Justice Central Files," National Archives

RG 67, NA   Record Group 67, "Records of the United

	States Fuel Administration," National Archives
RG 122, NA	Record Group 122, "Federal Trade Commission General Records," National Archives
RG 150, NA	Record Group 150, "Records of the National Bituminous Coal Commission," National Archives
Richberg Papers	Donald Richberg Papers, Manuscript Division, Library of Congress
Roosevelt Papers	Franklin D. Roosevelt Papers, Franklin D. Roosevelt Library, Hyde Park, New York
Slattery Papers	Harry Slattery Papers, Duke University Library, Durham, North Carolina
Smith Papers	Charles E. Smith Papers, University of West Virginia Library, Morgantown, West Virginia
UMW File	United Mine Workers of America File, AFL-CIO Library, Washington, D.C.
Wieck Papers	Edward A. Wieck Papers, Labor History Archive, Wayne State University, Detroit, Michigan
Wilson Papers	Woodrow Wilson Papers, Manuscript Division, Library of Congress

1

Perspective

AS A FOCUS for business-government relations from the
First World War through the Great Depression, the bituminous coal industry provides a complex subject. During the
war Congress created a Fuel Administration with plenary
power over the industry. Between the war and the New
Deal, governmental spokesmen proposed some sixty laws,
commissions, or investigations concerning the soft coal industry.[1] Union-sponsored legislation for soft coal in 1928
and 1932 and industry-structured price stabilization agencies helped shape the National Industrial Recovery Act. As
one of the major codes, soft coal occupied center stage during the early New Deal. Once the Supreme Court ended the
National Recovery Administration with the *Schechter* decision, the United Mine Workers of America convinced Franklin D. Roosevelt to endorse and Congress to pass the Guffey-Snyder Act of 1935. After the Supreme Court invalidated
that act in *Carter* v. *Carter Coal,* Congress passed another
special soft coal measure in the Guffey-Vinson Act of 1937.

Under Presidents Wilson, Harding, and Coolidge, national
interest in soft coal focused on consumers' problems of
shortages, strikes, and high prices. From the mid-1920's on,
lessened demand brought on by overdevelopment, competition from other energy sources, and economies in bituminous coal use reversed the situation. Hoover and Roosevelt

1. Glen Lawhon Parker, *The Coal Industry: A Study in Social Control* (Washington, D.C.: American Council on Public Affairs, 1940), chapter 5.

sought to restore profits and employment to a "sick" industry. Whether the issue was high or low prices, however,
from Wilson through Roosevelt soft coal played a significant
role in American business-government relations.

The politics of soft coal from the First World War through
the New Deal raises at least three kinds of questions. The
most important cluster of issues to historians concerns influence or control: whether government regulated the bituminous industry, was co-opted by the industry, or found some
middle ground. A second group treats the historical connection between the war experience and that of the New Deal. A
third set of issues involves the benefits, liabilities, and effectiveness of what was known as "industrial self-government."

The Question of Influence

A popular stereotype of business-government relations
sets the "public interest" against the business interests.
Progressives and New Dealers particularly stated the issues
of economic reform in those terms. Such a perspective has
informed the views of a number of scholars, particularly
those of a liberal persuasion, such as Charles Beard, Mat-
thew Josephson, Richard Hofstadter, Allan Nevins, and Ray
Ginger.[2] New Deal draftsman James M. Landis indeed
argued that New Deal economic regulation was built upon
a vision of the "public interest."[3] Historian Arthur M.
Schlesinger, Jr., constructed his magisterial *Age of Roosevelt* around the conflict between business and reformers.[4]
Franklin D. Roosevelt publicly inveighed against "economic royalists." John F. Kennedy remarked that his father
had told him that "all businessmen were sons-of-bitches,
but I never believed it till now."[5] President Jimmy Carter

2. In the order listed: *An Economic Interpretation of the Constitution of the
United States* (New York: The Macmillan Co., 1913); *The President Makers, 1896–
1919* (New York: Harcourt, Brace and Co., 1940); *The Age of Reform: From Bryan
to F.D.R.* (New York: Alfred A. Knopf, 1955); *Grover Cleveland: A Study in Cour-
age* (New York: Dodd, Mead, 1934); *Altgeld's America: The Lincoln Ideal Versus
Changing Realities* (New York: Funk & Wagnalls Co., 1958).

3. *The Administrative Process* (New Haven: Yale University Press, 1938).

4. *The Age of Roosevelt*, 3 vols. (Boston: Houghton-Mifflin, 1957–63).

5. Kennedy quoted in Arthur M. Schlesinger, Jr., *A Thousand Days: John F.
Kennedy in the White House* (Boston: Houghton-Mifflin, 1965), pp. 635–36.

blasted the large oil companies for wanting to "rip off" the American consumer.

Much recent scholarship has taken a different tack and has attempted to evaluate the actual political control exercised by the regulatory agency to determine the degree of influence the regulated industry has been able to exert on the regulatory agency. Employing this approach, scholars have found that the industrialists often captured or co-opted the regulatory agency or crucially influenced the legislation and appointees. Such a viewpoint to a greater or lesser degree is presented in the works of historians James Weinstein, Gabriel Kolko, Louis Galambos, Robert Wiebe, and Samuel Hays;[6] political scientists Grant McConnell, Marver Bernstein, Theodore Lowi, Earl Latham, Henry Kariel, and Robert Engler;[7] and sociologist C. Wright Mills.[8]

This argument has been set forth in great detail by Gabriel Kolko in *The Triumph of Conservatism* as well as *Railroads and Regulation, 1877–1916* and *Main Currents in Modern American History.*[9] "Federal economic regulation," he has written, "was generally designed by the regulated interest to meet its own end, and not those of the public or commonweal."[10] Thus, the so-called reforms of the Progressive Era can be described best as a "triumph" for conservative business leaders. During the 1930's, he stated, "politics becomes the means, once again, to engage more successfully in business."[11]

6. In the order listed: *The Corporate Ideal in the Liberal State: 1900–1918* (Boston: Beacon Press, 1968); *The Triumph of Conservatism* (Glencoe: The Free Press, 1963); *Competition and Cooperation: The Emergence of a National Trade Association* (Baltimore: Johns Hopkins University Press, 1966); *Businessmen and Reform: A Study of the Progressive Movement* (Cambridge, Mass.: Harvard University Press, 1962); *Conservation and the Gospel of Efficiency: The Progressive Conservation Movement, 1890–1920* (Cambridge, Mass.: Harvard University Press, 1959).

7. In the order listed: *Private Power and American Democracy* (New York: Alfred A. Knopf, 1966); *The Regulation of Businessmen* (New Haven: Yale University Press, 1954); *The End of Liberalism: Ideology, Policy, and the Crisis of Public Authority* (New York: W. W. Norton & Co., 1969); *The Group Basis of Politics: A Study of Basing-Point Legislation* (New York: Octagon Books, 1965); *The Decline of American Pluralism* (Stanford: Stanford University Press, 1961); *The Politics of Oil* (Chicago: University of Chicago Press, 1961).

8. *The Power Elite* (New York: Oxford University Press, 1956).

9. In the order listed: Princeton: Princeton University Press, 1965, and New York: Harper and Row, 1976.

10. *Triumph of Conservatism*, p. 59.

11. *Main Currents*, p. 143.

A third approach to the history of business-government relations emphasizes the structure of the industry and the interplay of interest groups which affect the regulatory process. Robert Wiebe's *Businessmen and Reform* portrayed the variety of internal geographic and economic divisions among businessmen in the same industry and showed how the rivalry among them affected the reforms of the Progressive Era. K. Austin Kerr, Lee Benson, Richard Vietor, Louis Galambos, William Graebner, and others have emphasized group interaction and shown the danger of treating regulation as a simplistic matter of the "people" defeating the "interests," or the regulated co-opting or controlling the regulators.[12]

For a variety of reasons, neither the "public-interest" model nor the "capture" model is as significant in understanding the politics of soft coal as is the structural approach. From the end of the nineteenth century the industry was chaotically disorganized. Because of its fragmentation it could not prevent or significantly alter the Lever Act, which placed it under strict wartime control. The industry's inability to stabilize itself helped insure declining economic fortunes for its owners during the 1920's. And because of its structural divisions, the soft coal industry could not use government to its own advantage during the 1930's, when it had government assistance. The UMW and the New Dealers did bring the industry to a moment of cohesion under the NRA code, but southern operators won a court injunction against enforcement of the code's wage provisions, and regional disputes were destroying the price stabilization program of NRA prior to the *Schechter* decision, which ended the experiment.

Because of soft coal's regional divisions and operator rival-

12. In the order listed: *American Railroad Politics, 1914–1920: Rates, Wages, and Efficiency* (Pittsburgh: University of Pittsburgh Press, 1968); *Merchants, Farmers, and Railroads: Railroad Regulation and New York Politics, 1850–1887* (Cambridge, Mass.: Harvard University Press, 1955); "Businessmen and the Political Economy: The Railroad Rate Controversy of 1905," *Journal of American History* 64 (June, 1977): 47–66; *Competition and Cooperation; Coal Mining Safety in the Progressive Period* (Lexington, Ky.: University of Kentucky Press, 1966). For a good survey of the literature on government regulation, see Thomas K. McCraw, "Regulation in America: A Review Article," *Business History Review* 49 (Summer, 1975): 159–83.

ries, the revived UMW—in league with some northern operators—emerged as the major power in the coal politics of
the 1930's. The UMW forced passage of two Guffey acts to
replace the NRA. But a Virginian's stockholder suit invalidated the first Guffey Act. Bureaucratic ineptness, regional
competition, and consumer pressures delayed the operation
of the second Guffey Act until the war-induced inflation
obviated it. But even though the UMW and the northern
operators became politically influential during the 1930's,
they failed to use government to stabilize the industry. Rivalry between the older northern fields and the newer,
lower-wage fields of the South gravely weakened all the
New Deal programs for soft coal. The politics of soft coal,
therefore, becomes understandable not by framing the
issues as the "people" versus the "interests," or by studying
the "control" of government over business or the reverse,
but by examining the industry's structural fragmentation and
fierce regional competition.[13]

The Analogue of War

The economic crises of war and depression provoked from
the American political system analogous responses. Both
Woodrow Wilson and Franklin D. Roosevelt abandoned the
traditional competitive economic order for a political economy which stressed cooperation among the men of business
and with those of government. War forced Wilson to jettison
his New Freedom; depression led Roosevelt to discard the
antitrust attitudes of the old progressive Democrats. In both
emergencies, rather than enforcing the antitrust laws, government fostered collusion among businessmen.

The Wilson and Roosevelt administrations established
similar bureaucracies. After painful delay, Wilson structured
a War Industries Board (WIB) to coordinate and increase
war production and to determine priorities for and allocation
of strategic materials. Congress created both a Fuel Administration and a Food Administration. Draftsmen modeled the

13. See McCraw, p. 181.

NRA in part after the WIB and its sister bureaucracies. The War Finance Corporation of the Wilson years led Hoover to create and Roosevelt to continue the Reconstruction Finance Corporation. Wilson's Council of National Defense, a committee of cabinet members which worked to prepare the nation for war, became the prototype for Roosevelt's Special Industrial Recovery Board. To win labor support and reduce strikes, Wilson set up the National War Labor Board; Roosevelt appointed the National Labor Board (NLB) and later the National Labor Relations Board. Special corporations used to build ships and control sugar prices during the war prefigured the New Deal's Commodity Credit Corporation. The Wartime Railroad Administration foreshadowed the office of Coordinator of Transportation.[14]

Various New Deal officials began their government service in wartime Washington. Charles F. Horner, who had led the Liberty Bond Drive, directed the NRA's publicity campaign. To head the NRA's Press Section and Speakers' Bureau, Horner brought back associates of the war years. Economist Leo Wolman served both the WIB and the NLB. George N. Peek, administrator of the New Deal Agricultural Adjustment Administration, had served in the WIB. Wayne P. Ellis, a veteran of the Fuel Administration, worked as a deputy administrator for the NRA's coal code. Former WIB dollar-a-year man Joseph P. Guffey introduced the New Deal's coal legislation.[15]

The overlord of the early NRA, General Hugh S. Johnson, labored during the war as both the organizer of the military draft and as chief of the General Staff's Purchase, Storage, and Traffic Division, where he collaborated with Bernard Baruch of the WIB. Johnson and Baruch became close friends and business associates; indeed, Johnson urged Roo-

14. John M. Blum, *Woodrow Wilson and the Politics of Morality* (Boston: Little, Brown and Co., 1956), p. 139; William E. Leuchtenburg, "The New Deal and the Analogue of War," in *Change and Continuity in Twentieth-Century America,* ed. John Braeman et al. (New York: Harper and Row, 1964), pp. 81–144; Gerald D. Nash, "Franklin D. Roosevelt and Labor: The World War I Origins of Early New Deal Policy," *Labor History* 1 (Winter, 1960): 39–52.

15. NRA Press Release 1864, Consolidated Files on Industries Governed by Approved Codes, Administration Members, RG 9, NA; Ray Sprigle, "Lord Guffey of Pennsylvania," *American Mercury* 39 (November, 1936): 280–83.

sevelt to appoint Baruch rather than himself to head NRA.[16] Easily the most visible and audible member of the New Deal apart from the president, the general often explained NRA in language reminiscent of the war years. NRA, he boomed to newsmen, "has behind it a greater idealism and necessity than that which made the draft effective."[17] He threatened that violators of NRA codes would meet the vigilante actions that dissuaded wartime motorists from driving on "gasless Sundays." Housewives, he proclaimed, " . . . will go over the top to as great a victory as the Argonne. It is zero hour for housewives. Their battle cry is 'buy now under the Blue Eagle!' "[18]

Assistant Secretary of the Navy Franklin D. Roosevelt, the most important veteran from the Wilson administration, had tried to create governmental war labor boards and to coordinate mobilization. He had even drafted a plan for an industrial economic council.[19] The NRA, Hugh Johnson remarked, was "peculiarly Franklin D. Roosevelt's own concept . . . his own particular property."[20] Although many diverse minds and experiences shaped the NRA, when it passed, Roosevelt recalled the Wilson years:

> I had part in the great cooperation of 1917 and 1918, and it is my faith that we can count on our industry once more to lift this new threat, and to do it without taking any advantage of the public trust that has this day been reposed without stint in the good faith, and high purpose, of American business. . . .
> As in the great crisis of the World War, it puts a whole people to the simple but vital test: "Must we go on in many groping

16. Frederic L. Paxon, *American Democracy and the World War*, vol. 1, *America at War* (Boston: Houghton-Mifflin, 1939), p. 221; Bernard M. Baruch, *Baruch: The Public Years* (New York: Pocket Books, 1962; originally published by Holt, Rinehart and Winston, 1960), pp. 236–37; Donald R. Richberg, *My Hero: The Indiscreet Memoirs of an Eventful but Unheroic Life* (New York: G. P. Putnam's Sons, 1954), p. 165.

17. Special Industrial Recovery Board, "Proceedings," June 26, 1933 (mimeographed copy), Columbia University School of Business Library; *New York Times*, July 30, 1933.

18. George Peek to Frank Walker, July 18, 1933, quoted in Leuchtenburg, p. 134; Hugh S. Johnson, *The Blue Eagle from Egg to Earth* (New York: Doubleday, Doran and Co., 1935), p. 264.

19. Nash, "Experiments in Industrial Mobilizations: W.I.B. and N.R.A.," *Mid-America* 45 (January, 1963): 160–61.

20. *New York Times*, July 21, 1933.

and disorganized separate units to defeat, or shall we move as one great team to victory?[21]

The 1920's, of course, provided a link between the two experiments in mobilizing the nation for united action, and the NRA and other New Deal programs were not taken directly from the war but grew out of many sources.[22] But the war did provide a model, and the New Dealers exploited it. But was it a valid model? Could the intense patriotic fervor of the war years be revived to combat the depression? The economic situations could not have been more dissimilar. Could techniques developed in a time of enormous economic demand be applied effectively in a stagnant economy? In short, did the war experiment provide the New Dealers with a useable past?

Once again, the structure of the bituminous industry played a significant role in determining its experiences under war and depression regulation. War-generated patriotism gave government momentary power to unify the coal industry and to restructure the industry's distribution and market system in 1918. Because government had wartime power, and because operators were making war profits, the industry cooperated with large-scale changes. A booming economy papered over the structural fragmentation of the industry. In depression, opposite economic conditions exacerbated North-South rivalries. Hard times heightened regional competition and worked to destroy the cooperation NRA sought. Without a war, New Dealers lacked the moral force the Wilsonians used to force changes. Southern operators in particular rebelled against the nationalizing tendencies of the New Deal programs for coal—particularly erosion of the South's wage differential. Even the Guffey acts could not overcome structural divisions. The New Dealers em-

21. Franklin D. Roosevelt, *The Public Papers and Addresses of Franklin D. Roosevelt*, 3 vols., ed. Samuel I. Rosenman (New York: Random House, 1938–50), vol. 2, p. 301.

22. Ellis Hawley, *The New Deal and the Problem of Monopoly* (Princeton: Princeton University Press, 1966); Robert F. Himmelberg, *The Origins of the National Recovery Administration* (New York: Fordham University Press, 1976); Ellis Hawley, "Herbert Hoover, the Commerce Secretariat, and the Vision of an 'Associative State,' 1921–1928," *Journal of American History* 61 (June, 1974): 116–40.

ployed the wartime rhetoric in vain. The war analogy proved to be false.

Industrial Self-Government

Personalities and institutions repeated themselves in war and depression administrations; so did the methods employed in shaping business-government relations during the two periods. In each era crisis caught public officials unprepared. In response, government sought to cooperate with business in a middle ground between laissez-faire and state control. Officials chose the technique of industrial self-regulation—a system in which industrialists kept control of the capital structure but cooperated with government agents to insure that corporate decisions did not violate the public interest.[23]

By the end of the war, the concept—if not the practice—of cooperation between government and industry flourished. Many were loath to abandon its advantages. The United States Chamber of Commerce proposed a peacetime cooperative scheme.[24] Impressed by the war-inspired teamwork between government and industry, Fuel Administrator Harry Garfield pressed Wilson to support a far-reaching plan which grouped the basic industries of the nation under a seven-man industrial cabinet.[25] Immediately following the armistice, Hugh Johnson began to lobby for a "high court of commerce . . . not a policeman, not a passive or negative agency, but a cooperator, an adjuster, a friend—in short, an agency set up on the theory of the war organizations, the Food, Fuel and War Industries Administrations." He recommended that a

23. William Leuchtenburg wrote, "Perhaps the outstanding characteristic of the war organization of industry was that it showed how to achieve massive government intervention without making any permanent alteration in the power of corporations," p. 129.

24. Robert A. Brady, *Business as a System of Power* (New York: Columbia University Press, 1943), fn. 19, p. 195.

25. "A Plan to Promote the Public Welfare by More Effective Cooperation Between the Government of the United States and Industry," in Harry A. Garfield to Woodrow Wilson, Feb. 26, 1919. See also Garfield to Wilson, March 27, 1919; Garfield to J. D. A. Morrow, March 18, 1919; and Garfield to Josephus Daniels, March 14, 1919, Records of Harry Garfield, Fuel Administrator, Records of the Executive Office, General Correspondence File, RG 67, NA.

statute endorse government-supervised trade associations to foster cooperation.[26] During his later days as NRA chief, Johnson boasted that his proposal of 1918 was "the N.R.A. idea complete. Its difficulties forecast and its essential logic stated 16 years ago."[27]

After the war, leading public figures supported the concept of industrial self-government. As secretary of commerce and as president, Herbert Hoover encouraged government cooperation with industry in a variety of activities ranging from seeking foreign markets to improving accounting practices. He particularly tried to help the bituminous industry stabilize itself.[28] In 1928, Federal Trade Commissioner Abram Myers stated that it was the "policy of the Federal Trade Commission to encourage self-government in industry and to avoid imposing governmental regulation whenever possible, because it realizes that those who are best acquainted with the peculiar problems of an industry are best equipped to govern it."[29] In 1930 Baruch urged a "common forum" for business "where problems requiring cooperation can be considered and acted upon with constructive, political sanction of government."[30]

New Dealers fell back on the ideal of industrial self-government because they wanted to avoid direct federal controls. In response to proposals by Alabama Senator Hugo Black, Baruch, Henry I. Harriman, president of the United States Chamber of Commerce, and General Electric's Ger-

26. Hugh S. Johnson's proposal, made as a report for the WIB in December, 1918, is reprinted in U.S., Congress, Senate, Committee on Finance, *Investigations of the National Recovery Administration, Hearings Pursuant to S. Res. 79*, 74th Cong., 1st sess., 1935, pp. 2409–11.

27. Ibid., p. 2411.

28. William A. Williams, *The Tragedy of American Diplomacy*, 2d ed. rev. (New York: Dell Publishing Co., 1962), pp. 109–15, 127–40; Herbert Hoover, *The Memoirs of Herbert Hoover*, vol. 2, *The Cabinet and the Presidency, 1920–1933* (New York: The Macmillan Co., 1952), pp. 64–84; Harris G. Warren, *Herbert Hoover and the Great Depression* (New York: Oxford University Press, 1959), pp. 27–29; Albert U. Romasco, *The Poverty of Abundance: Hoover, the Nation, The Depression* (New York: Oxford University Press, 1965), pp. 43–44; Ellis Hawley, "Secretary Hoover and the Bituminous Coal Problem, 1921–28," *Business History Review* 42 (Autumn, 1968): 247–70.

29. Myers's address quoted in U.S., National Recovery Administration, "Code Authorities and Their Part in the Administration of NIRA," NRA Work Materials No. 46 (mimeographed), Washington, D.C., 1935, p. 62.

30. Baruch's speech in Boston, May 1, 1930, quoted in Johnson, p. 156.

ard Swope, Roosevelt asked Congress in May, 1933, to con-
struct "machinery necessary for a great cooperative movement throughout all industry" to shorten hours, increase
wages, prevent unfair competition and excess production,
and reemploy workers.[31] He got the National Industrial Recovery Act. As administrator of the NRA, Hugh Johnson said
that he wanted "to avoid even the smallest semblance of
czarism. It is industrial self-government that I am interested
in."[32] NRA General Counsel Donald Richberg, later temporary chairman of the National Industrial Recovery Board,
declared that "in the National Recovery Administration we
have been seeking to carry forward our constitutional principles of self-government into a political-economic order. . . ."
The government, he said, would act "not as dictator, but as a
moderator—an umpire to insure fair play and the fulfill-
ment of public obligations."[33] Roosevelt proclaimed that
each code would "be administered by the industry itself,
with as little government interference and control as possible—the concept of industrial self-government."[34]

As a compromise between government control and laissez-
faire, the notion of self-government by industry had much to
recommend it, particularly in emergency situations. Owners
of capital retained essential control over their property while
government gained influence without having to seek specific
and possibly unconstitutional legislative action for individual
industries. Self-government further appeared democratic because it involved the concerned parties in governmental decisions affecting particular industries. Hugh Johnson boasted
that the "NRA itself is a democracy where industry, labor,
and consumers meet." The process insured that government
received knowledgeable business judgment on intricate eco-

31. Roosevelt quoted in NRA, "Code Authorities and Their Part," p. 59.
32. Johnson quoted in Schlesinger, *Age of Roosevelt*, vol. 2, *The Coming of the New Deal*, p. 110.
33. Address before the Institute of Public Affairs, Charlottesville, Va., July 9, 1934, quoted in NRA, "Code Authorities and Their Part," p. 62.
34. Roosevelt, vol. 2, p. 276. Following the demise of the NRA, a presidential committee analyzing the agency reported that self-government by industries had remained the central idea of the NRA. U.S., National Recovery Administration, Committee of Industrial Analysis, "Report" (mimeographed), Washington, D.C., 1937, pp. 17–18.

nomic problems and that government decisions would be more acceptable than if unilaterally determined.[35]

A central question about industrial self-government is, of course, whether it brought the results the advocates claimed it would. Leaving aside the question of influence, the issue is whether a system of industrial self-government was effective, for example, in streamlining distribution and increasing coal production during 1917 or in stabilizing prices and wages during NRA.

This book will show how the Wilsonians scrapped the concept of industrial self-government during the summer of 1917, even though the cooperative approach was improving production and distribution, because the progressives in the administration believed that the maximum prices the operators voluntarily set were too high. The Lever Act of 1917 supplanted industrial self-government. Later, the New Dealers turned back to self-government under the NRA in part because they believed that direct wage-and-price legislation would be declared unconstitutional. Self-government under the NRA stabilized prices momentarily. The revived UMW negotiated a contract which also stabilized wages at higher levels. But competition between regional code authorities, particularly North-South rivalries, weakened the New Deal's attempt at self-government. Once again structure proved more significant than a "public-interest" model or a "capture" model as a descriptive tool. Self-government for the fragmented coal industry was faltering even before the NRA was invalidated.

Related to the question of effectiveness of self-government is another question Theodore Lowi has raised about group-based legislation like the Guffey acts. He and other political scientists have argued that turning to the interest groups involved for legislative programs or administrative personnel tends inevitably to the atrophy of public control, "the maintenance of old and creation of new structures of privilege," and conservative ends.[36] The New Deal approaches to the coal industry—particularly the Guffey-

35. Johnson quoted in *New York Times*, November 7, 1933.
36. Lowi, p. 86.

Snyder Act of the liberal "second New Deal" Congress and the later Guffey-Vinson Act—must be examined in Lowi's terms, since they both attempted to protect the UMW and the interests of the northern operators and would have put the industry in an economic and bureaucratic straitjacket.

This book deals exclusively with the soft coal or bituminous industry. Anthracite, or hard coal, mined in eastern Pennsylvania, was monopolized in the 1870's by six railroad companies. Used almost exclusively for home heating, the higher-priced hard coal did not compete with bituminous, which was produced primarily for industrial uses. Dug in some five thousand mines in thirty-three states, bituminous coal did not lend itself to monopoly structure despite a merger movement at the turn of the century.[37]

The soft coal industry was divided into five major regions. One, known historically as the Central Competitive Field, encompassed the mines of Indiana, Illinois, western Pennsylvania, and Ohio. The northern Appalachian area included West Virginia, eastern Kentucky, part of Tennessee, and Maryland. Alabama, southern Tennessee, and Georgia formed the southern Appalachian region. Iowa, Missouri, Kansas, Oklahoma, and Arkansas made up still another regional group. The fifth region included the western states of Wyoming, Montana, New Mexico, Utah, Washington, the Dakotas, and Colorado.

The coal industry of this period is best understood by its regions. Even the largest firms like Pittsburgh Coal Company, Peabody Coal Company, and Consolidation Coal Company, which all owned mines in more than one state, still operated primarily in a single region. The operators' dependence on rail transportation with its local freight rates encouraged regional focus. The operators within one region also viewed themselves as members of producing subdistricts within parts of states, depending on the kinds of coal in their area, the mining conditions, the thickness of the coal

37. Joseph T. Lambie, *From Mine to Market: The History of Coal Transportation on the Norfolk and Western Railway* (New York: New York University Press, 1954), pp. 70–72; Ralph L. Nelson, *Merger Movements in American Industry, 1895–1956* (Princeton: Princeton University Press, 1959), p. 46.

seam, and railroad connections. Ohio itself, for instance, was divided into at least eight producing subdistricts, and these Ohio subdistricts were divided by owning company. Because of regional freight-rate differentials, however, local differences were often overridden by the group interests of the particular region.[38]

The crucial political and economic division in the soft coal industry pitted the northern operators of the older Central Competitive Field against the operators from the newer fields of the South. The Central Field had signed its first wage agreement with the UMW in 1898. Although collective bargaining collapsed for much of the Central Field in the 1920's, the operators there still paid much higher wages than did the non-union southern firms. State regulations on mine safety were also more advanced in the older northern fields.[39] Some freight rates also gave advantage to the South. All of these ingredients led to an unrelenting competition between northern and southern operators and to efforts by the North to employ government regulation to neutralize the competitive advantage of the South. The Central Field operators found an ally in the United Mine Workers of America, which wanted a national wage scale for its own reasons.

A complicated structure of coal markets provided some special advantage for certain regional coals but never dulled the competitive edge of the industry. Soft coal was shipped great distances to a variety of markets in which price was determined by a host of variables: the use to which the coal was put, the kind of coal, the freight rates involved, the method of transportation, the cost of production, the thermal value of the coal, its size, the reliability of the producer, his ability to secure railroad cars, and others. The major markets included all-rail shipment to the consumption centers of the industrial Midwest, where Pennsylvania, Ohio, and West Virginia coals competed with poorer-quality coal from Illinois and Indiana. West Virginia and Pennsylvania coal

38. Lambie, pp. 80–81.
39. William Graebner, "Federalism in the Progressive Era: A Structural Interpretation of Reform," *Journal of American History* 64 (September, 1977): 331–57, especially p. 353.

tended to dominate the tidewater markets of the Atlantic Coast—Charleston, Hampton Roads, Baltimore, Philadelphia, and New York City, from which coal moved by ship. Northern Appalachian and Central Field operators competed to sell "lake cargo coal," which went by rail to Lake Erie for water delivery to Detroit, Chicago, Milwaukee, and Duluth-Superior.[40]

Despite their regional, philosophic, union–non-union, and North–South divisions, however, the operators formed a key national energy industry. *Coal Age*, the major industry journal, attempted to speak for the industry as a whole. Although it specialized in engineering problems and emphasized technological matters, *Coal Age* repeatedly exhorted the industry to end its internecine struggles and stabilize itself. The industry's search for order led it to deal with government. The initial chapter of that relationship begins as the nation starts to mobilize for World War I.

40. Waldo E. Fisher and Charles M. James, *Minimum Price Fixing in the Bituminous Coal Industry* (Princeton, Princeton University Press, 1955), chapter 5; William Graebner, "Great Expectations: The Search for Order in Bituminous Coal, 1890–1917," *Business History Review* 48 (Spring, 1974): 50; Lambie, plate XV, shows lake markets.

2

From Cooperation to Control

BETWEEN 1914 AND 1918 the Wilsonians developed a series of bureaucratic structures to harness American industries to the war objectives. Managed for the most part by dollar-a-year business advisors like Bernard M. Baruch, head of the War Industries Board, these agencies commonly relied upon cooperation and gentlemen's agreements rather than sanctions. Many industries, in fact, brought to the war bureaucracy their own skilled spokesmen and trade organizations developed during the Progressive Era's search for order. Needing the cooperation of big business to fight the war, the Wilsonians relaxed earlier antitrust efforts and moved toward accommodation with large corporate enterprise.[1]

In the WIB, conflict of interest became less important than efficiency, coordination, and production. The nominee from a particular industry often became the government's dollar-a-year man in dealing with that trade. Prices the government paid for steel, oil, copper, hides and skins, aluminum, and cotton goods were negotiated in friendly bargaining between industry leaders and government purchasers and allowed ample profits. Both during the war and after, the steel industry under the leadership of Judge Elbert

1. Robert A. Wiebe, *The Search for Order, 1877–1920* (New York: Hill and Wang, 1967), pp. 293–302; Robert D. Cuff, *The War Industries Board: Business-Government Relations during World War I* (Baltimore: Johns Hopkins University Press, 1973), pp. 3–11, 148–49, 167–73, 218; Mark Sullivan, *Our Times: The United States, 1900–1925*, vol. 5, *Over Here* (New York: Charles Scribner's Sons, 1933), pp. 489–91.

Gary, chairman of United States Steel, "shaped federal policy to meet its own desires." Auto makers successfully maintained auto production and resisted WIB demands to shift to war production. Indeed, well-organized industries dominated business-government relations during the war.[2]

During the summer of 1917, under the leadership of operator Francis S. Peabody of Chicago and at the request of Wilson administration spokesmen, the coal industry formed the Committee on Coal Production (CCP) to help mobilize the industry for war. The CCP attempted to get operators to streamline distribution, increase production, and hold down skyrocketing prices. Earlier befriended by Federal Trade Commissioner Edward N. Hurley, the coal men were encouraged by Secretary of the Interior Franklin K. Lane to set maximum prices.

But other Wilsonians of an antitrust frame of mind distrusted the coal men. Attorney General Thomas Gregory, Secretary of the Navy Josephus Daniels, Secretary of War Newton D. Baker, and President Wilson saw the CCP as a manifestation of the "coal trust," a conspiracy to raise prices. The administration repudiated the CCP maximum prices and disgraced its leaders. Wilson then set maximum coal prices himself and signed legislation which placed the coal industry under sweeping federal control.

Ironically, federal control over coal came not because the industry was monopolistic but because it was not. Coal industrialists fell under federal control in part at least because they had been unable to cooperate in the prewar years. Ignoring its fragmented structure and perhaps confusing it with the oligopolistic hard coal industry, its enemies within the administration denounced the soft coal industry as a "trust." But the bituminous coal industry had failed before the war to develop programs to stabilize its production, mar-

2. Melvin I. Urofsky, *Big Steel and the Wilson Administration: A Study in Business-Government Relations* (Columbus: Ohio State University Press, 1969); and Robert D. Cuff and Melvin I. Urofsky, "The Steel Industry and Price-Fixing during World War I," *Business History Review* 39 (Autumn, 1970): 291–306; Gerald D. Nash, *United States Oil Policy, 1890–1964* (Pittsburgh: University of Pittsburgh Press, 1968), pp. 29–36; Bernard M. Baruch, *Baruch: The Public Years* (New York: Pocket Books, 1962; originally published by Holt, Rinehart and Winston, 1960), pp. 62–64; Cuff, pp. 58–70, 128–29, 227–40.

keting, or prices. The industries which had rationalized their affairs found ways to manipulate the Wilsonian war bureaucracy. Soft coal remained disorganized, highly competitive, inefficient, and wasteful. When America entered the war, the industry had no national trade association, monopoly power, or leadership.

Between 1880 and 1917 various operators—primarily from the Central Competitive Field—had tried to fashion marketing associations, sales agencies, trade associations, and other stabilizing structures. They had failed. Still, when coal prices rose in 1916–1917, antitrust progressives blamed a mythical "coal trust," characterized the industry's attempt to cooperate with the emerging Wilsonian war mobilization program in 1917 as a conspiracy in restraint of trade, and demanded full government control of coal. Because skyrocketing demand and transportation snarls had raised coal prices, the industry found few defenders. In short, the fragmented bituminous coal industry's attempt to cooperate with the war mobilization program backfired and led to extensive federal control under the Lever Act of 1917.

Stabilization Fails, 1880–1917

The bituminous industry in America began on an optimistic note. During the industrial growth of the late nineteenth and early twentieth centuries, coal production surged. From an average annual output of 30 million tons in the 1870's, the industry grew each decade to World War I—including even the depressed 1890's, when coal production jumped by 92 percent—until 1913 production hit 478 million tons, as shown in Fig. 1. Even in its boom years, however, the economic structure of the industry portended that one day it would be classified as a "sick" industry.[3]

Before the First World War, the bituminous industry held its own in the energy market. In 1918, as Fig. 2 shows, soft coal supplied the nation with nearly 70 percent of its en-

3. Waldo E. Fisher and Charles M. James, *Minimum Price Fixing in the Bituminous Coal Industry* (Princeton: Princeton University Press, 1955), chapter 1.

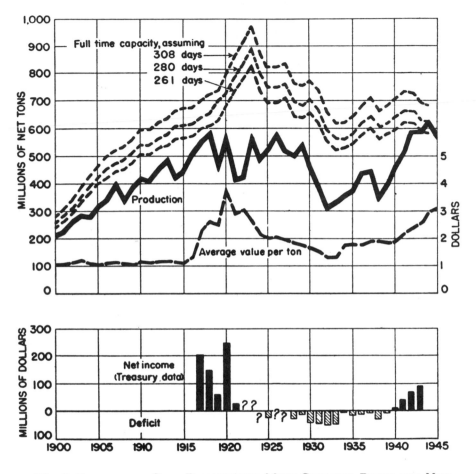

Fig. 1. Bituminous Coal Production, Mine Capacity, Price, and Net
Income or Deficit, 1900–1945.

Source: U.S. Bureau of Mines, *Minerals Yearbook*, 1945.

ergy. But, spurred by technological developments in diesel-
powered ships, automobiles, trucks, and oil heating equipment, petroleum producers reduced coal's share of the energy market to 50 percent by the Great Depression.[4]

4. Nash, pp. 4–6.

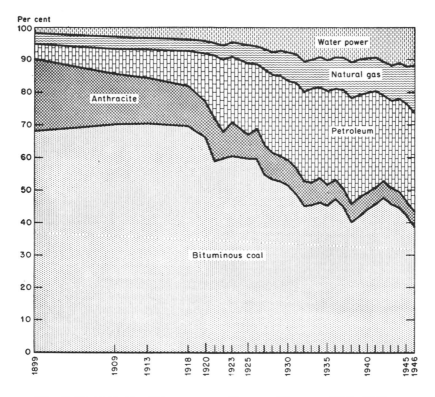

Fig. 2. BITUMINOUS COAL IN THE ENERGY MARKET, 1899–1945.
Source: U.S. Bureau of Mines, *Minerals Yearbook*, 1945.

For a number of reasons, from the 1880's onward, however, the industry's capacity to produce exceeded the demand for coal. Availability of labor and seemingly inexhaustible reserves of a "ready-made" product that required little technology to produce encouraged entrepreneurs to overdevelop the industry. The railroads often charged low ton-mile rates for long hauls from new locations and thus spurred mine development. Seasonal demand and inability to store coal further led to excess capacity. Because the geographical areas containing coal deposits were widely separated, operator cooperation in attempting to curtail the industry's elastic supply proved impossible.[5]

Although aggregate demand for coal kept increasing until the mid-1920's, short-run demand for coal remained inelastic. Lowered prices did not increase quantity demanded. For the primary users of soft coal—the railroads, the public utilities, and the steel mills—whose use was governed more by the rate of industrial activity than by price, lower average coal prices would not increase total coal sales. The f.o.b. mine cost of coal was simply too small a part of their total costs.[6]

In fact, the large industrial purchasers of coal, often described as monopsonists or oligopsonists, were in a better bargaining position than the coal operators. One spokesman seeking governmental protection for coal prices asserted that "the coal operator merely meets the price the purchasing agent of the railroad, utility, manufacturing plant, or retailer tells the producer he will pay."[7] This argument can be overdrawn, because metallurgical coals and other particular coals with high thermal efficiency or other special characteristics gave certain sellers a bargaining advantage. Favorable freight rates, geographical location, and contracts stabilized markets for particular coals.[8] Also, as steel companies and utilities integrated vertically, their "captive mines" removed up to one-quarter of yearly bituminous output from the open market. Yet despite the influence of these variables, the bulk of American coal was sold in buyers' markets.[9]

5. Carroll Lawrence Christenson, *Economic Redevelopment in Bituminous Coal* (Cambridge, Mass.: Harvard University Press, 1962), pp. 220–21; and Clair Wilcox, *Public Policies toward Business* (Chicago: Richard D. Irwin, Inc., 1955), p. 466. See particularly William Graebner, "Great Expectations: The Search for Order in Bituminous coal, 1890–1917," *Business History Review* 48 (Spring, 1974): 49–73.

6. Reed Moyer, *Competition in the Midwestern Coal Industry* (Cambridge, Mass.: Harvard University Press, 1964), pp. 53–61; Wilcox, p. 465; U.S., Temporary National Economic Committee, *Competition and Monopoly in American Industry*, Monograph No. 21 (Washington, D.C.: Government Printing Office, 1940), p. 25.

7. K. C. Adams, director of research, United Mine Workers of America, testimony in U.S., Congress, Senate, Committee on Mines and Mining, *To Create a Bituminous Coal Commission: Hearings before a Subcommittee*, 72d Cong., 1st sess., 1932, p. 793.

8. Graebner, p. 50; Christensen, pp. 18–20, 86–94, 152, 193; Moyer, pp. 160–61.

9. Jacob Schmookler, "The Bituminous Coal Industry," in *The Structure of American Industry*, ed. Walter Adams (New York: The Macmillan Co., 1954), pp. 76–100.

This condition has led to assertions that before World War II the bituminous industry was an example of perfect competition. The major governmental investigation of monopoly power in American business, conducted by the Temporary National Economic Committee of the 1930's, stated: "It is clear that no one mine and no small group of mines controls a share of the national output large enough to enable it to determine or substantially to influence the price of coal." Plagued by intense price competition before 1914, the industry would meet it again after 1924 when aggregate demand slacked off and competition from other fuels increased.[10]

Despite its competitive structure, however, antitrust critics persisted in characterizing the soft coal industry as the "coal trust." The anthracite strike of 1902 inflamed public sentiment hostile towards arrogant anthracite "coal operators." In their self-righteous refusal to arbitrate differences in a dispute with the United Mine Workers of America, the railroad presidents and their mine managers had indeed been perfect targets for public ridicule. But George Baer's autocratic comments about the managers of capital being "the Christian men to whom God in His infinite wisdom has given the control of property interests in this country" were words of a railroad president. Yet the image of unscrupulous "coal barons" was fixed and applied to soft coal operators.[11]

In 1913, South Carolina Democratic Senator "Pitchfork Ben" Tillman launched an investigation of the "power and influence of the so-called Coal Trust." He aimed not at producer monopoly, but at shipper control over coal production. Tillman alleged that the Southern Railway Company had spurred the opening of the Virginia, Tennessee, and Kentucky coal fields by promising direct rail connections to tidewater markets "until the mining capacity is probably ten times greater than is needed to supply the demand in the consuming territory to which these mines are restricted. . . ."

10. Temporary National Economic Committee, *Competition and Monopoly*, p. 24; see also Blaine Davis, "The Marketing Problems of Bituminous Coal," *Harvard Business Review* 11 (November, 1932): 99; and Walton Hamilton and Helen R. Wright, *The Case of Bituminous Coal* (New York: The Macmillan Co., 1925), p. 42; Moyer, pp. 67–68, 200–205.
11. *Literary Digest* 24 (1902): 824.

Then "the Big Interests who may defy the law and laugh at the Interstate Commerce Commission" got control of the Southern Railway Company and ended the construction of the tidewater connections in favor of the coal trade on other northern railroads that they managed. Even this fighter for the coal industrialists of the South against the northern railroad managers described his enemies as "the big coal interests."[12]

Faced with waste and inefficiency, operators sought to rationalize the industry between 1890 and 1914. They tried various means: expanding foreign markets; equalizing production costs through uniform state laws on safety and accident compensation—even agreeing to try the stabilizing influence of union wages; trying to persuade the Interstate Commerce Commission to equalize transportation costs; attempting to create a national operator association; establishing regional selling agencies; negotiating price agreements among themselves; and merging coal firms.[13]

Yet none of these programs met the problem of excess capacity. Increased exports at the turn of the century proved inadequate. Uniform state legislation failed when delegates from Ohio and West Virginia refused to participate in the Uniform Mining Laws Conference of 1916. The ICC offered no substantive relief from regionally oriented railroad rates.

Joint sales agencies which marketed coal for their members and paid the operators the average price of the coal sold worked briefly. The operators in Ohio's Hocking Valley, for example, created a successful sales agency during the depression of the 1890's. For regions with particular coals or in less competitive markets, sales agencies performed a service. But they often dissolved when an individual company decided to seek its own advantage. Similarly, price-fixing associations like the Southern Coal Association,

12. B. R. Tillman to W. W. Finley, president, Southern Railway Company, May 22, 1913; Tillman to Daniel Willard, president, Baltimore & Ohio Railway Company, June 5, 1914; Tillman to J. C. McReynolds, attorney general of the United States, May 29, 1914, and July 15, 1914; draft of Senate Resolution 291 on "the power and influence of the so-called Coal Trust," all in Classified Subject Files, Correspondence, File 60-187-13, RG 60, NA.
13. Graebner, pp. 49, 71.

created in 1895 for the operators in Tennessee, Alabama, and Kentucky, succumbed when Tennessee coal men began to disregard the price schedules. Government prosecution and conviction of a West Virginia selling agency under the Sherman Act in 1899 dampened enthusiasm for both sales agencies and price-fixing agreements.[14]

Turn-of-the-century operators tried mergers. The creation of Pittsburgh Coal Company, Monongahela River Consolidated Coal and Coke Company, H. C. Frick Company, and Consolidation Coal Company did not, however, produce monopolistic control. In fact, no firm gained control of more than 3 percent of coal production before the war; the large firms competed rather than cooperated with each other, and independents continued to insure competitive pricing and production. A potential scheme to create a giant Midwestern consolidation modeled after U.S. Steel and Standard Oil and under the leadership of J. P. Morgan collapsed in the antitrust climate following the *Standard Oil* decision of 1911.[15]

The national trade association movement in coal failed as well. In 1909, operators from Illinois, the state which would be at the center of efforts to stabilize the industry throughout this study, began negotiations to create the American Federation of Coal Operators. But West Virginia's coal industrialists, who anticipated that membership in a national association would force them to recognize the United Mine Workers of America, rebuffed the midwesterners. The regional rivalries which would plague the industry from the war to the New Deal were beginning to be heard. Competitive instincts led Pittsburgh operators to remain aloof. Split into three regional trade associations, Ohio operators found they could not unite statewide, much less enter a national association. Only Illinois, Indiana, and western Kentucky joined the American Federation of Coal Operators.[16]

Except in the Midwest, the union movement also failed to stabilize production costs in the prewar period. Local miners' associations decried the "insane competition" and "in-

14. Ibid., pp. 49–72.
15. Ibid., pp. 59–61.
16. Ibid., pp. 55–56.

discriminate cutting of prices" resulting from the railroad growth of the late nineteenth century. The operators for the most part, however, resisted the workers' plans for interstate cooperation. One wealthy Ohio coal man did join with unionists in 1886 to organize a conference in Columbus, Ohio, to "stabilize the industry." This February 23, 1886, conference, attended by representatives of miners and operators from Pennsylvania, Ohio, Indiana, Illinois, West Virginia, and Maryland, set wage scales for their districts as a way of equalizing competitive conditions. The midwestern miners had begun a long struggle to stabilize the industry under union auspices which would climax in the New Deal.[17]

Although the depression of the 1890's and withdrawals by some operators destroyed this initial, union-led effort, in 1890 the Knights of Labor District Assembly No. 135 and the National Progressive Union of Miners and Mine Laborers merged to produce the United Mine Workers of America (UMW). The fledgling UMW led a strike in 1897 that paralyzed the northern fields but failed in the South. A landmark in mining history, this strike brought operators in Pennsylvania, Ohio, Indiana, and Illinois into the first of many wage agreements of the Central Competitive Field with the UMW. This 1898 agreement established the eight-hour day, a standard wage for men who worked on a daily basis, and a standard tonnage rate with local wage differentials. The Central Competitive Field operators recognized the UMW and created the structure for wage stabilization in that region based on what the participants called the principle of "competitive equality."[18]

17. Philip A. Taft, *Organized Labor in American History* (New York: Harper and Row, 1964), pp. 166–68; Graebner, p. 55.
18. John L. Lewis, *The Miners Fight for American Standards* (Indianapolis: United Mine Workers of America, 1927), p. 42; McAlister Coleman, *Men and Coal* (New York: Farrar & Rinehart, 1943), pp. 49–50; Robert Wiebe, *Businessmen and Reform: A Study of the Progressive Movement* (Chicago: Quadrangle Books; also published by Harvard University Press, 1962), p. 161; Van Bittner, "Wages in Bituminous Coal Mines as Viewed by the Miners," *Annals of the American Academy* 3 (January, 1924): 40; K. Austin Kerr, "Labor-Management Cooperation: An 1897 Case," *Pennsylvania Magazine of History and Biography* 49 (January, 1975): 45–71.

The opening of the coal fields of the South soon undermined the Central Field Agreement and created the key North-South economic schism within the industry. Better coals, easier mining conditions, and advantageous ton-mile railroad rates combined to encourage development of the southern fields. The absence of the union there made southern wage scales lower and southern coals more competitive in energy markets. Strikes shut down Ohio's mines during much of 1914 and 1915, and in combination with recession and southern competition they pushed one-fourth of Illinois tonnage into bankruptcy in those years. Central field operators realized that unless they could find a way to equalize competitive conditions between the North and the South, they faced further losses to the newer mines there.[19]

After the financial panic of 1907, the Central Field operators began to turn increasingly to the federal government for aid. One coal executive advocated socialistic control by government. But most advocates of federal action sought creation of a federal commission to oversee and permit stabilizing legal combinations, contracts, and trade agreements. The coal men had difficulty convincing congressmen that their industry was unusually competitive, and the Federal Trade Commission was definitely not the agency they desired. They failed to have James Callbreath, secretary of the American Mining Congress, appointed to the FTC, but meetings with Interior Secretary Franklin K. Lane gave hope that the FTC might support coal trade agreements.[20]

Over the winter of 1914–1915, Callbreath and other midwesterners pressed the FTC, the Department of Justice, and President Woodrow Wilson for the right to create sales agencies and open-price associations to help stabilize markets for Indiana and Illinois coal. They argued that intense competition there drove prices down, led to waste of natural resources, and brought unemployment and inefficiencies in mining. In a series of letters and exhibits the midwesterners

19. F. G. Tyron, "Effect of Competitive Conditions on Labor Relations," *Annals of the American Academy* 3 (June, 1924): 85–89.

20. The first issue of *Coal Age* carried the plan of John A. Jones, president of Pittsburgh-Buffalo Coal Company, to have government set both prices and wages in the industry. *Coal Age* 1 (October 14, 1911): 15; Graebner, pp. 64–67.

contended that reasonable combinations to protect price levels and reduce competition were in the public interest. The FTC and the president met with them, studied their exhibits and legal arguments, but refused to sanction the proposals. "Congress," the FTC acting chairman wrote, "has given the Commission no power or jurisdiction to pass upon proposed combinations and advise as to their legality."[21]

That same year a committee of Central Field operators worked with the UMW to amend the Federal Trade Commission Act to allow trade combinations. A committee of Charles M. Moderwell, a Franklin County, Illinois, operator; Charles Keith, a Kansas City, Missouri, coal man; and Carl Scholz, president of the American Mining Congress, got UMW President John P. White to draw the American Federation of Labor (AFL) and the United States Chamber of Commerce into the plan. The operators' committee wrote AFL President Samuel Gompers that "some form of cooperation is imperative if this industry is to be saved from bankruptcy; a stop put to the wicked waste of a vital and natural resource; some protection given to the capital invested; and the lives and wages of the worker properly safeguarded." A conference on December 18, 1915, between the operators and the AFL leaders and one between Gompers and John H. Fahey, president of the chamber, on December 29 broadened the effort to reach a proposal which would have sought FTC protection for "reasonable cooperation" in all industries. It came to naught.[22]

Without administration approval, but with the implied support of FTC Vice Chairman Edward N. Hurley, sales

21. Callbreath to Wilson, April 1, 1915, with enclosure; Wilson to Callbreath, April 26, 1915, Reel 292, Wilson Papers. Carl Scholz, president, American Mining Congress, to Callbreath, n.d.; Scholz to Joseph E. Davies, chairman, Federal Trade Commission, March 25, 1915; Callbreath to Wilson, April 21, 1915; Scholz to Franklin K. Lane, secretary of the interior, May 15, 1915, in File 8508-10-1-1, RG 122, NA. Sidney A. Hale to F. E. Berquist, November 14, 1935, in Division I, Reports, Consolidated Files on Industries Governed by Approved Codes (bituminous coal), RG 9, NA; Wiebe, p. 84; *Coal Age* 3 (April 17, 1913): 571–72; 7 (May 8, 1915): 806–807; (May 20, 1915): 866–67.

22. L. C. Boyle to Joseph E. Davies, January 22, 1916; John H. Fahey to Charles S. Keith, September 7, 1915; C. M. Moderwell, Carl Scholz, and Charles S. Keith to Chamber of Commerce of the United States of America, December 2, 1915; John P. White to Charles S. Keith, April 17, 1916, in Federal Trade Commission General Records, 1914–21, File 8149-13, RG 122, NA.

agencies and open-price associations sprang up in the Midwest. A successful businessman and Chicago Democrat who traveled extensively speaking to various trade groups to encourage business rationalization, Hurley became the industry's friend in government. He met with the operators and assured them that their plans were legal as long as they did not result in "unreasonable prices." It was, *Coal Age* noted, a "pleasing experience to hear an accredited representative of the Government admit that the coal industry is in the mire, needs to be systematized and made profitable." The initial Illinois organization, the Franklin County Coal Operators' Association, divided the Franklin County territory into zones, each overseen by a commissioner. By reporting to the commissioner all prices and tonnage figures on coal shipped to a particular zone and studying his reports, the members could ascertain whether or not their prices were lower than the averages and adjust them. Because no operator was required to raise his prices and because no presale price fixing was involved, the operators hoped that the Justice Department would not prosecute. The successful concept spread to Indiana, Ohio, and St. Louis, Missouri.[23]

At the meeting of the American Mining Congress in mid-November, 1916, Hurley, who had been directly influenced by Arthur J. Eddy's concepts of the "new competition," advocated his pet scheme of uniform cost accounting as an initial step in efficient operation. The FTC, he suggested, would help the industry, "but the industry must take itself in hand and work out its own salvation." If uniform cost accounting alone did not suffice to reduce unrestricted competition, however, "if the well-being of a great industry demands a forward step in national policy, let us not be afraid to take it." He personally would endorse, he said, "any measure of government regulation that may be found necessary in order to effect a real remedy."[24]

23. *Coal Age* 9 (June 17, 1916): 1072; (January 8, 1916): 78; (May 27, 1916): 947; 10 (September 9, 1916): 434.

24. Hurley's speech is in ibid. 10 (November 25, 1916): 887–89. Hurley went on in 1917 to publish *Awakening of Business* (Garden City, N.Y., 1917), which encouraged cooperation along the lines he was suggesting to the coal men. *Coal Age* 10 (December 23, 1916): 1047.

The FTC had begun to study the production and distribution of bituminous coal as directed by a resolution representing Central Field ideals and introduced by Congressman Henry T. Rainey, Democrat of Illinois. Unfair methods of competition had demoralized the industry by encouraging inefficiencies in mining, argued the resolution, citing the loss of two hundred million tons of coal and increased unemployment and part-time employment among the miners. These conditions, noted the resolution, "vitally touch the interests and welfare of the public as a whole." *Coal Age* judged the resolution "wholly meritorious." Hurley envisioned the study as a way to generate industrial cooperation with the FTC.[25]

Yet Hurley could not speak for the entire FTC or for the Wilson administration. Other FTC commissioners were becoming increasingly suspicious of the coal-price associations. One FTC document noted that because "the product is largely sold on contracts for future delivery," the reporting of past prices "might be of considerable influence in preventing active price competition." The industry had found its friend in the federal bureaucracy, and Congress had commissioned the FTC to study the industry, but operator Charles Keith had little hope that the trade commission would help the fragmented coal industry. Worse, Justice Department officials had begun preparation of suits involving antitrust violations by both coal dealers and two coal trade associations that investigators believed were "running into illegal practices" like price fixing and withholding unsold coal from open markets.[26]

Thus, by early 1917 the industry could show little for its stabilization efforts. Foreign markets had not absorbed the excess productive capacity. The open-price associations were under attack, and the joint selling agencies, mergers, and price-fixing arrangements either had failed or had been

25. The text of the resolution is in *Coal Age* 10 (November 18, 1916): 851.

26. "Memorandum Regarding Various Statements and Plans Submitted by Illinois and Indiana Coal Operators," Federal Trade Commission General Records, 1914–21, file 8508-10-2-1, RG 122, NA; George Anderson, U.S. attorney, to Thomas Gregory, U.S. attorney general, January 6, 1917, Straight Numerical File No. 181092-98½, RG 60, NA.

put aside as possibly illegal. Southern production and defections by particular midwestern operator associations had undercut the wage stability of the UMW contracts in the Central Competitive Field. Amendments modifying antitrust legislation had been dropped. No national trade association had been created. Although Central Field operators had been among the few coal men who had sought to organize the industry, *Coal Age* argued that even they showed "a lack of leadership and a state of incapacity which is little less than criminal" compared with the rest of the "commercial, financial, and manufacturing world [which] is replete with associations doing effective work for their respective lines of business in this country." On the eve of the Great War, the bituminous coal industry remained essentially competitive, disorganized, and leaderless.[27]

The Impact of War

War revolutionized the industry's economic conditions. In 1916, aggregate demand for coal skyrocketed, and as prices rose, incentives to organized action among coal operators collapsed. Price increases and labor shortages replaced price cutting and unemployment. Almost overnight the war revived the previously demoralized industry. Allied orders for iron, steel, and munitions flooded into America. Coal, the source of nearly three-fourths of the nation's energy requirements in this period, suddenly seemed in short supply. Between the recession of 1913–14 and the American entry into the war in April, 1917, yearly coal output mounted by 30 percent, or 130 million tons. In 1914 the average weighted price for all bituminous in America stood at $1.16. By 1916 it had risen slightly, and new contracts for the coal year pushed the average in April to $2.30.[28]

27. *Coal Age* 8 (September 18, 1915): 153; 10 (November 25, 1916): 801.
28. Paul W. Garrett, "History of Government Control of Prices," in *History of Prices during the War*, ed. Wesley C. Mitchell (Washington, D.C.: Government Printing Office, 1920), pp. 153, 170; U.S., Federal Trade Commission (henceforth cited as FTC), *Report on Anthracite and Bituminous Coal* (Washington, D.C.: Government Printing Office, 1917), p. 15. Between 1913 and 1917 average wholesale prices rose 70 percent. George Soule, *Prosperity Decade* (New York: Harper

The railroads and other large consumers had contracts that pulled the average down.[29] But smaller consumers who bought coal at market prices rather than under annual contracts faced price increases of between 200 and 400 percent.[30] Public attention riveted on these spot prices. In 1915, Pittsburgh run-of-mine coal sold for less than one dollar in the open market; by the spring of 1917 it commanded over five dollars. Secretary of the Navy Josephus Daniels noted this price in his diary in May as he prepared to fight for lower prices for the navy's fuel.[31] Shortages, caused in large part by lack of railroad cars to move coal from the mines, generated panics. Speculators moved into the market. One man allegedly went into the mining camps with suitcases full of cash. He bought the entire output of a mine at the end of one day and auctioned it off the next morning.[32]

By the time the country had entered the war in April, rising coal prices had killed interest in proposals to stabilize the industry and had reactivated the turn-of-the-century image of the industrialists as unscrupulous barons. The activities of sharp coal dealers and speculators reawakened the public's fears of the "coal trust." William B. Colver, a Wilsonian progressive on the FTC, turned the FTC study of overdevelopment in coal into an investigation of shortages and price increases.[33] In a pointed contrast to Hurley, Colver wrote Wilson that the bituminous operators had "at no time . . . as a class risen above the most sordid conception of

and Row, 1968; originally published by Holt, Rinehart and Winston, 1947), p. 56. The average weighted prices of 1,365 commodities increased 56 percent between June, 1913, and March, 1917. Cuff, p. 94. Charles Richard Van Hise, *Conservation and Regulation in the United States during the World War* (Washington, D.C.: Government Printing Office, 1917), pp. 31–32; George Cushing, "Ending the Coal Dilemma," *Atlantic Monthly* 122 (November, 1918): 590.

29. David I. Wing, "Cost, Prices, and Profits of the Bituminous Coal Industry," *American Economic Review* 11 (March, 1921): 77.

30. FTC, *Anthracite and Bituminous Coal*, p. 18.

31. Josephus Daniels, *The Cabinet Diaries of Josephus Daniels*, ed. E. David Cronon (Lincoln: University of Nebraska Press, 1963), p. 151; U.S, Fuel Administration (henceforth cited as FA), *Final Report of the United States Fuel Administrator, 1917–1919* (Washington, D.C.: Government Printing Office, 1921), p. 20.

32. U.S., Congress, Senate, Committee on Interstate Commerce, *Price Regulation of Coal and Other Commodities, Hearing on S. 2354 and S.J. Res. 77*, 65th Cong., 1st sess., 1917, pp. 55–56.

33. FTC, *Anthracite and Bituminous Coal*, pp. 13–14.

their relation to the public and the nation and to the war emergency."[34]

Always cool to the cooperative activities of the coal industrialists, the Justice Department won the first round in its cases against the coal sales agencies. On April 9, 1917, a grand jury indicted twenty-one corporations and eighteen officials for conspiring to increase prices of bunker coal. In another case the government won an indictment against eight individuals and thirty-five corporate defendants of the Smokeless Coal Operators Association of West Virginia for price fixing. The assistant attorney general admitted that the smokeless coals were "the best steaming coals mined," yet he believed that the price jump by the association's market committee to three dollars a ton was "the most extortionate that has occurred in the coal industry. . . ."[35]

The government and the coal industry had. arrived at a crossroads. Justice Department officials and William Colver could argue that war required that the "coal barons" be brought to heel and that government directly fix prices and regulate the industry in the public interest. Supported by Secretary Franklin K. Lane's Department of the Interior, coal men could counter that federal price fixing or control was unnecessary, given the available excess productive capacity.[36] Once transportation bottlenecks were opened, demand would draw out production and reduce prices. Coal men knew that demand had more than doubled between 1900 and 1913 while prices held steady. But this steady growth over time differed from a sudden shift in the industry's demand curve. New mines took time to open, and even if more coal could be mined, how could the existing rail network and supply of cars get hundreds of thousands of

34. William B. Colver to Wilson, August 16, 1917, Subject File, Fuel Administration, Box 126, Garfield Papers; Joseph Tumulty to Senator Francis Newlands, February 2, 1917, Reel 308, Wilson Papers.

35. *New York Times*, April 10, June 28, July 7, 1917; Frank M. Swacker to G. Carroll Todd, assistant to the attorney general, February 24, 1917, including "Memorandum Regarding Alleged Combination in Restraint of Trade in Bituminous Coal . . . ," Classified Subject Files, Correspondence, File No. 60-187-16, RG 60, NA.

36. On Lane's appointment to the Department of the Interior, see Arthur Wallworth, *Woodrow Wilson* (Baltimore: Penguin Books, 1958), book I, p. 272.

additional tons to markets? The "coal problem" was in large part a transportation problem.

Secretary of the Navy Josephus Daniels, a spokesman for the populist traditions of the South, felt that the government should fix both government and consumer prices of coal, oil, and copper as had England and France. In May, 1917, he informed the coal operators that in wartime a "man who made more than normal profits was little less than a traitor." If the operators would not voluntarily agree on a special price for navy coal, he later stated, then the government "ought to commandeer all the coal output to prevent the profiteering." Operators, he felt, "owned coal only as trustees for the country.[37]

War had suddenly reversed old arguments. It had been as trustees of the country's natural resources that the operators appealed for validation of programs to stabilize the industry and stop waste and inefficiency.[38] Although antitrusters like Daniels had been suspicious of the coal men all along, the operators had been right. The industry had been wastefully competitive in a natural resources sense, and its highly competitive nature had led coal operators to seek government support for stabilization. In 1917 government officials were condemning an essentially fragmented and leaderless industry as if it were a monopoly.

Cooperation Backfires

The war pressed the government to coordinate industrial mobilization. In August, 1916, in response to a flood of proposals to inventory industrial resources, Congress created a subcommittee of the cabinet as a Council of National Defense (CND). As chairman, Wilson appointed his new secretary of war, Newton D. Baker. Representing their departments on the CND were Josephus Daniels, secretary of the navy; Franklin K. Lane, interior; David F. Houston, agricul-

37. Daniels, *Cabinet Diaries*, May 23, 1916, notes of a speech; May 25, 1917; June 16, 1917; and May 29, 1918, pp. 156–57, 169, 308.

38. "Memorandum Regarding Various Statements and Plans Submitted by Illinois and Indiana Coal Operators," RG 122, NA.

ture; William B. Wilson, labor; and William C. Redfield, commerce.[39]

Baker, the diminutive, pipe-smoking former reform mayor of Cleveland, argued for voluntary solutions to the problems of mobilization. A progressive Democrat, Baker had fought for home rule for Cleveland; he opposed all federal, bureaucratic proposals. He believed in traditional American institutions and hoped that the war emergency would not disturb the customary American reliance on local, individual effort. Although he distrusted the coal operators, he also opposed direct federal price fixing in favor of prices set in conferences between government and producers. Cooperation, Baker wrote to Wilson, is the "only alternative" to government seizure of the mines, which "neither Congress nor the public are ready to sanction."[40] Composed of busy cabinet members, Baker's CND had to rely for most of its work on the volunteer labors of its Advisory Commission. To this Advisory Commission, Wilson appointed leaders of various realms of enterprise—most of them men who had been urging preparedness action since 1915.[41]

Creation of the CND with its Advisory Commission of what became known as "dollar-a-year" men both followed the logic of the situation and set the entire mobilization effort in a framework of informal agreements. Wilson brought private spokesmen representing various interest groups into official status as advisors to the cabinet officials responsible for mobilization. The administration sought government and industry cooperation achieved by business advisors. In Wilson's words, the new agencies would open up "a new and direct channel of communication and cooperation between business and scientific men and all depart-

39. Cuff, chapter 1; see also Frederick Palmer, *Newton D. Baker*, 2 vols. (New York: Dodd, Mead & Co., 1931), vol. 1, pp. 48–50; William F. Willoughby, *Government Organization in War Time and After* (New York: D. Appleton-Century Co., 1919), pp. 11–13.

40. Cuff, p. 64; Urofsky, p. 207; Daniel R. Beaver, "Newton D. Baker and the Genesis of the War Industries Board, 1917–1919," *Journal of American History* 52 (June, 1965): 44–45; Baker to Wilson, May 28, 1917, Box 4, Baker Papers.

41. Sullivan, vol. 5, pp. 377–78; Cuff, pp. 12–14, 18–19, 27–30, 39; Palmer, vol. 1, p. 61; Willoughby, p. 13; Baruch, p. 23.

ments of the Government."[42] Whether this technique would work for a leaderless, disorganized industry like coal remained to be seen.

As the nation shifted from preparedness to war that winter, the CND and its business advisors on the Advisory Commission moved toward establishing informal price agreements with major suppliers. To determine standards for munitions production, the CND in February, 1917, created a Munitions Standards Board, which the CND broadened into the General Munitions Board a month later. As chairman of the munitions board, Frank Scott appointed a price committee. Under Baker's mandate "to act on questions involving the determination of fair and just prices of munitions and related supplies" to be purchased by the government, the price committee issued a price report on May 3, 1917. The committee sought "cooperation of the great owning and operating interests" to achieve "sufficient price regulation." Governmental threats of retaliation against noncompliance might be employed, said the report, but the key to the price problem lay in cooperation between business and government.[43]

This report coincided with the views of both Baker and Bernard M. Baruch, who as raw materials chief was moving to center stage in the mobilization program. Although disagreeing over techniques, both men stressed cooperative agreements among men of goodwill. His familiarity with the titans of industry made the tall, dignified Baruch particularly fear the impact of government price fixing. The former genius of Wall Street shared his profession's hostilities to government regulation. "Dr. Facts," as Wilson called Baruch, understood the workings of American business. He expected to be able to use his personal touch with business leaders to get cooperation in holding down prices. Baker and Baruch opposed the inordinate complexity of a government-instituted, cost-plus system of fixed prices. Their coop-

42. Wilson quoted in Cuff, p. 40. Baker to Wilson, September 18, 1916; Walter S. Gifford to Baker, May 26, 1917, and an undated, unsigned memorandum in Reel 238, Wilson Papers; see also Baker to Wilson, May 28, 1917, Baker Papers.

43. Palmer, vol. 1, p. 130; Cuff and Urofsky, p. 293; Willoughby, pp. 70–71; Cuff, p. 96. On the relations between industry and government, see Cuff, chapter 3.

erative approach would result, perhaps, in higher prices.[44] But in the early months of the war the president and top government officers agreed that prices were secondary to output.

Under CND supervision, various industries formed cooperative, industrial self-government committees. Although organized on a voluntary basis, and often bypassed by individual businessmen as they sought contracts from government, these committees formed a significant reservoir of information and talent. Gifford later testified that as a practical matter the Advisory Commission had to delegate much work to the committees. Baruch tended to deal directly with the individual leaders in particular industries. Yet he, too, had praise for the efficiency of the committee work during the early phases of the war. Reciprocally, the committees became power bases for businessmen who wanted to insure business—as opposed to military—control over mobilization. The business leaders on the cooperative committees sought to insure their predominance in the emerging system.[45]

The CND created a broad-based, fifteen-member Committee on Coal Production, representing operators, miners, and consumers. Dominated by nine leading operators from the major producing areas chosen largely on recommendation of Secretary Lane, the CCP also included John Mitchell, former president of the United Mine Workers of America, and representatives from the Bureau of the Mines, the U.S. Geological Service, the by-product coke business, and the consumers of New England, where transportation difficulties had produced severe shortages. The CND appointed Mr. Francis S. Peabody, a Chicagoan noted both for his successful presidency of the Peabody Coal Company there and for his ability to organize his fellow operators, chairman of the CCP. A man of humanitarian impulses, Peabody was warmly endorsed by the industry's leading journal, *Coal Age,* which doubted that "a more unselfish collection of

44. Baruch, p. 86; Cuff, pp. 55, 58, 100; Daniels, *Cabinet Diaries*, May 28, 1917, p. 158; Baker to Wilson, May 28, 1917, Box 4, Baker Papers.
45. Cuff, pp. 59–85; Paul A. C. Koistinen, "The 'Industrial-Military Complex' in Historical Perspective: World War I," *Business History Review* 41 (Winter, 1967): 378–403.

men could have been brought together for the purpose in hand." Creation of the Lane-Peabody committee, the journal noted, meant that "the patriotism of the coal-mining industry will make it unnecessary for Congress to pass any radical legislation to secure the necessary results."[46]

On May 9, 1917, the CCP met with Gifford, Baruch, and Secretary Lane, the industry's chief administration contact. Baruch outlined the committee's tasks. Lane provided the patriotic incentive: "War was once a question of feeding the soldiers. Now it is an industrial game, the foundation of which is coal." Without coal, Lane stressed, "the war cannot be carried on." The CCP's job, announced Lane, was to stimulate production and, by working with the railroads, to increase the movement of coal to markets. The operators were assigned regions, where they were to organize other operators in a great cooperative effort to boost output and move coal. Although exhorting the operators to cooperate, Lane warned, "It is a question of cooperation or compulsion; but instead of taking over the mines and the great industrial plants, we are asking you to serve on this fuel board." *Coal Age* averred that the attitude of officials in the nation's capital is "one of hope" that under the "advisory guidance of this new committee," coal prices "will quickly regulate themselves."[47]

The Peabody committee appealed to operators and miners to step up production, to load cars more efficiently, and to coordinate transportation.[48] Following more meetings, the members fanned out across the nation to encourge their regions to increase output. They organized state committees of operators, formed consumer groups, created the Tidewater Coal Exchange to expedite shipments to the allies and the

46. *Coal Age* 11 (May 5, 1917): 782; (May 12, 1917): 817–18; form letter dated July 3, 1917, in Subject File: Navy, Coal, Daniels Papers; United Mine Workers of America to Wilson, May 23, 1917, Reel 292, Wilson Papers; *New Republic* 11 (June 9, 1917): 146; Cushing, p. 593.

47. *Coal Age* 11 (May 19, 1917): 867; Lane quoted in *Coal Age* 11 (May 12, 1917): 818, 835; testimony of Mr. Harry M. Taylor, member of the CCP, in Senate Committee on Interstate Commerce, *Price Regulation*, p. 85; testimony of Charles Moderwell, member of CCP, in Senate Committee on Interstate Commerce, *Price Regulation*, p. 29.

48. *Coal Age* 11 (May 19, 1917): 867–68.

Northeast, appointed committees of miners to discourage strikes, and designated publicity spokesmen. In hundreds of meetings throughout the nation, producers and railroad men discussed bottlenecks in transportation, reconsignment of coal cars, and the abuses of jobbers. In the West, operators' committees renegotiated the union contract upward ten cents a ton to avoid a labor conflict. The larger consumers absorbed the increase by agreeing to higher coal prices. Within weeks, the Committee on Coal Production created a gigantic network. As one member, Mr. Henry N. Taylor of Missouri's Central Coal and Coke, said, this committee "goes back to the zones, and from the zones to the cities, and from the cities to the coal mines and to the consumer's bin."[49]

The first weeks of the voluntary program of self-regulation brought mixed results. Production increased between April and May by over five million tons—a 12 percent gain, but typical of the rather normal increase following contract negotiations at the beginning of the new coal year on April 1. Still, production was running ahead of 1916 by eight or nine million tons monthly. On May 26, 1917, Walter S. Gifford, the director of the CND's Advisory Commission, wrote the secretary of war that the Lane-Peabody committee "has already accomplished marked, definite results." But production gains failed to reduce spot prices. Between April and May spot coal prices jumped from $3.00 to $3.72, and they rose another nickel the following month. Participating coal men reported localized production successes, but transportation bottlenecks continued to plague New England. Increasing production was not bringing prices down.[50]

On June 19, 1917, the Federal Trade Commission dropped a bomb on the CCP. It published a report on anthracite and bituminous coal that challenged the entire cooperative program. The commission's report of exactly a

49. Testimony in Senate Committee on Interstate Commerce, *Price Regulation,* pp. 86–89; George W. Reed, secretary of the CCP, to Todd, July 5, 1917, and Todd to Reed, July 5, 1917, Classified Subject Files, No. 18109-A32, RG 60, NA.

50. Walter S. Gifford to Baker, May 26, 1917, and May 28, 1917, Reel 238, Wilson Papers; testimony of Charles M. Moderwell in Senate Committee on Interstate Commerce, *Price Regulation,* p. 25; Garrett, pp. 153, 170.

month before, recommending improvements in transportation, labor, and storage conditions, had been received as "lucid and reliable" by the industry. The earlier report had hinted that the government might have to set prices and regulate distribution, but these proposals drew scant notice. The second report, advocating the pooling of all coal production under government controls, stunned the industry. The government would pay operators for their coals on a cost-plus basis and then resell all coal. To insure efficient delivery of coal, the government would coordinate and control the entire water and rail transportation system. It was a "gigantic scheme," said the *New York Times. Coal Age* damned it as "extraordinary and revolutionary."[51]

At this juncture two progressive senators, Atlee Pomerene, Democrat of Ohio, and Albert B. Cummins, antitrust Republican from Iowa, introduced proposals empowering the government to fix coal prices. Cummins wished to use the Interstate Commerce Commission's power to forbid railroads from purchasing coal at prices higher than those of June 30, 1916. Like trade commissioner Colver, and in apparent cooperation with him, Pomerene sought to have Wilson act through the FTC. In the Senate Interstate Commerce Committee's hearings on the two drafts at the end of June, 1917, Chairman Francis J. Newlands, Democrat from Nevada, argued that he reluctantly had come to the conclusion that excessive demand for basic products required that government "act and act decisively" to prevent inflation from raising the cost of government "and at the same time raise prices of necessaries beyond the reach of the masses of the people." If government could conscript citizens, he said, it could conscript the materials of war at a just price.[52]

51. *Coal Age* 11 (June 2, 1917): 947–49; (June 30, 1917): 1120; *New York Times,* June 21, 1917, FTC, *Anthracite and Bituminous Coal,* pp. 13–20; William J. Harris, chairman of the FTC, to Wilson, April 7, 1917; Wilson to Harris, April 9, 1917; Commissioner Fort made a preliminary report on May 3; Wilson sent a copy to Newton Baker, May 4, 1917, and expressed his gratitude to Fort the same day, Reel 292, Wilson Papers; see also Wilson to Indiana Governor J. P. Goodrich, June 18, 1917, Reel 292, Wilson Papers.

52. Senate Committee on Interstate Commerce, *Price Regulation,* pp. 3–7; Cushing, p. 594.

The coal industry and the Peabody committee faced another powerful antagonist in Navy Secretary Josephus Daniels, who sought to purchase the navy's 1918 requirements—approximately 1.7 million tons—at $1.72 per ton. Using a prewar navy fuel market survey, Daniels stubbornly forced operators negotiating through the Peabody committee to come down from their offer of $2.95 a ton to $2.335 pending an FTC investigation of costs. He jokingly told Peabody and others in one negotiation hearing, "If I had the power, I would pay you to run your business. . . . If anybody slacked working any mines, I would court-martial him." The navy's obdurate stance irritated the industrialists, but it gave notice that some Washington officials would not accept even the semblance of profiteering in coal.[53]

Thus chastened, various representatives from the Peabody committee went to testify before the Senate Committee on Interstate Commerce about the FTC pooling plan. Some reconsidered their earlier emphasis on increasing production to lower prices. They admitted that some operators had gouged the public. Although midwesterner Charles M. Moderwell of the fuel committee still championed stimulating production to lower prices, he recognized that the government might turn price fixing over to the FTC. He realized as well that the fuel committee faced a conflict-of-interest problem if they took over price fixing themselves. Indeed, he and other fuel committee spokesmen argued that the CCP feared even discussing price fixing, thinking such actions violated the antitrust laws.[54]

To forestall government price-fixing legislation, the CCP

53. Garrett, p. 155; testimony of Naval Commander Hilton in U.S., Congress, Senate, Committee on Manufacturers, *Publication of Production and Profits in Coal, Hearings on S. 4828,* 66th Cong., 3rd sess., 1921, p. 1315; "Minutes of Coal Conferences, May 29, 1917," Subject File: Navy, Coal, p. 42, Daniels Papers; "Coal Bidders" memorandum in Subject File: Navy, Coal, Daniels Papers; Daniels to Wilson, June 18, 1917; Wilson to Daniels, June 19, 1917, Special Correspondence, Woodrow Wilson, Daniels Papers; *Coal Age* 11 (June 16, 1917): 1041; Urofsky, p. 121.

54. Colver testimony, pp. 213–22; Moderwell testimony pp. 24–64; Taylor testimony, p. 100, all in Senate Committee on Interstate Commerce, *Price Regulation. Coal Age* 12 (July 7, 1917): 76.

summoned over four hundred operators to meet in Washington on June 26–28, 1917. The operators discussed production, distribution, and rising prices. Federal Trade Commissioner John F. Fort warned them that the president had directed the FTC to produce a study of coal costs. But Fort said the government would not prosecute them under the antitrust laws if they cooperated to lower prices. Lane appealed to their patriotism. "The success of the country depends on you," he exhorted, "and you are responsible, just the same as the soldiers in the trenches, Pershing in France, or the President in the White House." He described how George Washington had risked his fortune in the Revolutionary War. "Is there a man who will say to me, 'you can send your boy to France, while I stay here to coin his blood into dollars?' " The operators answered with loud applause.[55]

Encouraged, the operators determined to lower prices voluntarily in advance of the FTC study. They delegated seven men for each coal-producing state to confer with the CCP and with its approval "fix immediately a fair and reasonable price on coal f.o.b. cars at mines in each district." By June 28 they made the major decisions for most areas. The CCP announced tentative maximums, pending the FTC cost study. The operators provisionally fixed the price for Pennsylvania mine-run coal—a bench mark for the industry—at three dollars a ton. Other coals got their usual differentials. *Coal Age* judged the prices a "sweeping reduction." Because the CCP took responsibility for the decision, the operators felt they had skirted the antitrust laws. Lane praised the new prices as "very large reductions" from the market rates.[56] He declared to the mass meeting:

> Gentlemen, this is a very novel proceeding. I think I am within the fact when I say that no such hearing or gathering

55. *New York Times,* June 27, 1917; Tumulty to Senator Francis Newlands, February 23, 1917, Reel 308, Wilson Papers; Baker to Wilson, June 30, 1917 (not sent), Box 4, Baker Papers.

56. *Coal Age* 11 (June 30, 1917): 1124; 12 (July 7, 1917): 33; 15 (January 30, 1919): 227; *New York Times,* June 28–29, 1917; Cushing, p. 594; *New York Times,* June 29, 1917; Lane testimony in U.S., Congress, Senate, Committee on Manufactures, *Shortage of Coal: Hearings before the Subcommittee of the Committee on Manufactures Pursuant to S. Res. 163,* 65th Cong., 2d sess., 1918, pp. 538–39.

as this has ever been held in the United States before, or perhaps in the world. You are, I hope, pioneers in a good movement.... You have said to the American people that within your power, exercising your judgment, protecting yourselves, you will not be oblivious to the rights of those whom you serve; you will, within your power, protect them. That is the spirit that makes for the success of our country. And if all the industries of the United States will have that same spirit, there will be no question as to our ability to mobilize the resources of this country and carry this war to a successful conclusion.... We are the greatest business Nation on earth, and therefore we must look to the businessmen of the country to lead our people in spirit. And I think that the word that comes out from this gathering will be an inspiration to the people of the country.[57]

The entire price list, far below recent spot quotations, was enunciated probationally. Wilson's nephew, J. Wilson Howe, whom Wilson believed knew "a good deal" about the coal men, wrote the president that the CCP had "the necessary organization ready and are prepared to handle any phase of the business and in any manner that the government desired." They were "willing to have the Federal Trades [*sic*] Commission fix prices, for they want nothing more than a fair profit at any time. . . . But," he warned, "they feel that their efforts to cooperate should be appreciated and receive some recognition."[58]

By attempting to forestall government price-fixing legislation with their own voluntary maximums, the operators blundered twice. To consumers their prices seemed too high. And to progressives of an antitrust or antibusiness persuasion, they resembled a revived "coal trust." In fixing temporary maximum prices, pending the results of the FTC cost study, the operators inadvertently destroyed the cooperative relationship with wartime government. Those seek-

57. Lane's speech reprinted in *Congressional Record*, 65th Cong., 1st sess., 1917, vol. 55, part 5, p. 5318.
58. Department of the Interior press release, June 28, 1917, in *Congressional Record*, 65th Cong., 1st sess., 1917, vol. 55, part 5; Howe to Wilson, August 13, 1917, Box 91, Garfield Papers; see also the correspondence between Lane and George Peabody reprinted in *Congressional Record*, 65th Cong., 1st sess., 1917, vol. 55, part 5, p. 5317.

ing direct government control over the industry soon swept the field.

Passage of the Lever Act

The three-dollar price agreement jolted the Wilson administration. On the evening it was announced, Baker summoned Lane, Attorney General Thomas W. Gregory, Secretary of Labor Wilson, and Commissioner Fort to his office. Gregory fumed that the operators had merely reduced levels to the same prices for which the Justice Department was prosecuting other operators. Fort felt that the high prices were needed to draw out more tonnage. Increased production, he reasoned, would soon pull prices below the operators' maximums. Baker said he would follow whatever course Wilson desired on the matter.[59]

Furious over the operators' maximums, Wilson told Baker and Gregory that he "planned to repudiate, publicly, the price fixed." He shared Gregory's sentiments that the government could not both endorse the CCP three-dollar maximum and prosecute the Appalachian operators for an alleged minimum at a similar level. Wilson wrote Baker that he awaited the FTC investigation of coal costs as a basis on which he "might be instrumental in bringing about an understanding which would be in every way fair and reasonable." By encouraging the operators to fix their own prices, wrote Wilson, the CND and the FTC were working "at cross purposes." The coal prices, moreover, were "clearly too high." Wilson asked to be kept informed of any future price-fixing conferences.[60]

Baker suggested that he, not Wilson, repudiate the prices agreed to under Lane's supervision. In a lengthy statement, approved by Wilson, to the Advisory Commission's director

59. Daniel R. Beaver, *Newton D. Baker and the American War Effort, 1917–1919* (Lincoln: University of Nebraska Press, 1966), p. 65; Ray Stannard Baker, *Woodrow Wilson: Life and Letters,* 8 vols. (Garden City, N.Y.: Doubleday, Page, 1927, and Doubleday, Doran, 1927–39), vol. 7, p. 137; *New Republic* 11 (July 7, 1917): 258.

60. Wilson quoted in Baker, vol. 7, p. 138; Wilson to Baker, personal and confidential, June 29, 1917, Box 4, Baker Papers; Beaver, p. 65; Daniels, pp. 169–70; Baker, vol. 7, pp. 125, 135–36.

Walter S. Gifford, Baker blasted the CCP prices on two grounds. The CND, he asserted, "has no legal power and claims no legal power, either to fix the price of coal, or to fix a maximum price for coal or any other product." Based on initial data from the FTC study, Baker damned the three-dollar maximum as "an exorbitant, unjust, and oppressive price." Lane, who came to the brink of resigning, decided to remain in the cabinet. Shortly after the incident Lane and Baker lunched publicly together and discounted their differences.[61]

The CCP price agreement had embittered consumers and antitrust progressives. Now the administration's repudiation of the agreement jarred the coal community. Many conference participants justly felt that Lane had encouraged them to do precisely what Baker, as secretary of war and chairman of the CND, attacked them for. In obvious disgust, Peabody stormed out of Washington. His secretary, George Reed, told anxious coal operators to maintain the voluntary price reductions. Peabody wrote Baker later and charged that the administration had encouraged consumers to stop ordering coal for the winter. Peabody predicted that the disruption of the trade caused by Baker's letter would mean shortages and that "hardship and suffering will result."[62]

The CCP maximums, in fact, brought the CND's cooperative approach into question. High officials wanted total reorganization of the CND and creation of a central purchasing agency headed by Baruch.[63] Wilson set out to strengthen the administration's power over consumer prices, as Daniels had recommended. Congressman Asbury F. Lever's food

61. The Baker letter is in Garrett, pp. 158–59; Wilson to Fort, July 2, 1917, in Baker, vol. 7, pp. 140–41; Willoughby, p. 295; *Coal Age* 12 (July 14, 1917): 76, 158–59; Palmer, vol. 1, pp. 372–73; *New York Times*, July 3, 1917; Cuff, pp. 105–106; Wilson to Baker, July 2, 1917, Baker Papers, Box 4; Franklin K. Lane to Robert Lansing, May 14, 1921, in Franklin K. Lane, *The Letters of Franklin K. Lane, Personal and Political*, ed. Anne W. Lane and Louise H. Wall (Boston: Houghton-Mifflin, 1922).

62. *Coal Age* 12 (July 7, 1917): 33; (July 14, 1917): 76; Peabody to Walter S. Gifford, July 13, 1917, in Garrett, p. 159, fn. 2; Cuff and Urofsky, pp. 293–304; Cuff, pp. 93–98.

63. See comments by Senator Boise Penrose in *Congressional Record*, 65th Cong., 1st sess., 1917, vol. 55, part 5, p. 5317. Baker to Wilson, July 3, 1917; Wilson to Baker, July 17, 1917; Baker to Wilson, August 9, 1917; and Wilson to Baker, August 16, 1917, all in Box 4, Baker Papers; *New Republic* 11 (July 7, 1917): 266–68; *New York Times*, July 2, 1917.

control bill proposed to regulate production "of food and other necessities of life." Ohio Senator Atlee Pomerene tossed a price-fixing measure for coal into the Senate hopper and met with Wilson on federal control of coal. Wilson backed Pomerene's bill as an amendment to Lever's bill. Drawn apparently from Colver's FTC report, the amendment empowered the president to fix coal prices either directly and/or by requiring the operators to sell through a government agency, presumably the FTC.[64]

Pomerene bluntly attacked the operators for attempting to create a fictitious shortage of coal by side-tracking coal-laden railroad cars. The "coal barons," he alleged, were making a profit of at least two dollars a ton on coal. If allowed to continue, these profits would grow to over a billion dollars a year. German submarines were sinking American ships. He vowed that the coal industrialists should not be allowed to "submarine all the manufacturing interests of my own State, and of the country."[65]

Pomerene and other senators received letters from small businessmen, merchants, and homeowners detailing their problems in obtaining cheap coal. Senator John K. Shields from Tennessee reasoned that the high prices alone must "convince almost anyone that there is some sort of trust existing between [sic] the operators." Mississippi Senator James K. Vardaman asserted that the operators were "taking advantage of this, the most extraordinary emergency in the history of the nation to rob and plunder the people." He excoriated the "plutocratic patriots" who "fill to plethora their lean coffers with ill-gotten gold."[66]

The coal industry became a scapegoat. Although steel,

64. *New York Times*, July 4, 1917; Baker, vol. 7, p. 121; Thomas J. Walsh to Wilson, June 28, 1917, Reel 294, Wilson Papers; *Congressional Record*, 65th Cong., 1st sess., 1917, vol. 55, part 5, p. 5313, and part 4, p. 3558; undated memorandum for Harry A. Garfield from H. D. Nims, Records of Harry Garfield, Fuel Administrator, Records of the Executive Office, Interoffice Memoranda, RG 67, NA.

65. *Congressional Record*, 65th Cong., 1st sess., 1917, vol. 55, part 5, p. 4692; *New York Times*, July 6, 1917; U.S. Coal Commission, *Report*, 68th Cong., 2d sess., 1925, pp. 2517–24.

66. *Congressional Record*, 65th Cong., 1st sess., 1917, vol. 55, part 5, pp. 4684–96.

copper, iron, and aluminum prices had all skyrocketed, the Senate voted down an amendment to include them in a government price-fixing plan, 50–27. Oil prices were not regulated until 1918, and then voluntarily. The Senate rejected proposals to empower the president to designate particular articles for price control. Republican Senator Boise Penrose of Pennsylvania, a well-established friend of the coal operators, attempted to explain that speculators and middlemen—not any plot by the operators—caused the swift increase in spot coal prices. The Senate did not accept his argument. Instead, it adopted the Pomerene amendment to the food bill, 72–12. Penrose groaned that "there seems to be some kind of mysterious subterranean agreement for mutual protection and defense which the coal operators and the farmers apparently failed to get into."[67]

Throughout the spring and early summer, a jury in New York had been considering the government's case against the Appalachian operators who were alleged to have fixed the price of smokeless coal at three dollars. The prosecutors admitted that the defense had piled up great amounts of evidence on the competitive nature of the industry, including the data that smokeless coal was then selling at five dollars a ton and up. But they contended that the smokeless operators' action in agreeing on a three-dollar price, "which the defendants were free to observe or not as they chose," was an undue restraint of trade. Amid all the public furor over the CCP price of three dollars a ton, the jury found the defendants not guilty. The court dismissed a parallel case against other operators. In post-trial questioning, the jurors stated that they believed that the operators had conspired but that their price, "due to increased demand," had little effect on the market. Assistant Attorney General G. Carroll Todd wrote the prosecutor in New York that the time had

67. Cuff, p. 87; *New York Times*, July 14, 20, 21, 22, 1917; Joseph P. Guffey, *Seventy Years on the Red-Fire Wagon* (Lebanon, Pa.: published by the author, 1952), pp. 54–57; Nash, pp. 20–34; *Congressional Record*, 65th Cong., 1st sess., 1917, vol. 55, part 5, p. 5316. The negative voters: Brandegee, Chamberlain, Colt, France, Knox, Lodge, McLean, Penrose, Sutherland, Wadsworth, Warren, and Weeks.

come to either stiffen up the Sherman Act or admit "that it is a broken reed."[68]

By continuing the cooperative relationships with other industries, the administration placed coal in a unique position. Alleged collusion in aluminum and shoes sparked a proposed conflict-of-interest amendment to the food control bill that any dollar-a-year advisor could not participate in a contract in which he might have a financial interest. Stating that the government needed the "cooperation and the active cooperation, of the men who are in actual control of the great business enterprises of the country," Wilson fought the amendment. His pending reorganization of the mobilization program would, he promised, protect against direct involvement of business advisors in contracts in which they had an interest. But it would allow these men to provide the government the information and cooperation it needed. Atlee Pomerene's substitute proposal that allowed an advisor to recommend contracts as long as he first disclosed in writing any interest he might have in it pleased Wilson and passed the Senate, 53–17.[69]

The administration defeated another Senate amendment proposing to create a joint congressional committee to oversee the mobilization program; Congress ratified the Lever bill, and Wilson signed it on August 10. Under Section 25 the president now had power to establish maximum prices for coal "wherever and whenever sold." He could, if he chose, "require any or all operators" to sell coal to the government through an agency set up by him for that purpose.[70]

The passage of the Lever Act thrust the coal industry into a period of uncertainty. Following the signing, Wilson

68. *New York Times*, July 7, 13, 1917. John Lord O'Brian, special assistant to the attorney general, to G. Carroll Todd, March 28, 1917; Robert P. Stephenson to O'Brian, July 13, 1917 (two letters); Isaac R. Oeland to Attorney General Thomas W. Gregory, July 13, 1917; Todd to O'Brian, July 16, 1917; Oeland to Todd, July 17, 1917, all in Classified Subject Files, Correspondence, File 60-187-16, RG 60, NA.
69. Wilson quoted in Cuff, p. 107; Cuff, pp. 108–111; Urofsky, p. 165; *Congressional Record*, 65th Cong. 1st sess., 1917, vol. 55, part 5, pp. 4596ff.; *New York Times*, July 14, 17, 1917; Cuff, "Woodrow Wilson and Business-Government Relations during World War I," *Review of Politics* 31 (July, 1969): 392–407.
70. Lever Act, 40 Stat. 276ff.; *New York Times*, July 30, 1917.

leaked his intention to employ the "big stick" policy on industrialists—particularly coal and steel men—who refused to cooperate with government. The FTC announced that it expected to roll back anthracite prices and warned bituminous producers that they might expect similar action. Representing various public utilities at an FTC hearing, Clifford Thorne urged the government to lower coal prices and end alleged extortionate profits. Prominent officials across the country urged Wilson to take control of the coal industry. The FTC finished its study on coal production costs. On the afternoon of August 20, 1917, Wilson walked to the commission offices and spent three hours going over cost sheets with FTC officials.[71]

The next day Wilson fixed prices himself. The president cut the CCP prices by one-third. He set Pittsburgh run-of-mine coal at two dollars a ton, with customary differentials for other districts. Contracts were maintained. He would reconsider his prices, said the president, "when the whole method of administering the fuel supplies of the country shall have been satisfactorily organized and put into operation." The coal situation was "a tough question," he wrote Fort. "I am by no means confident that I have struck the right judgment."[72]

Massive operator protest poured into Washington. *Coal Age* argued that prices would have soon fallen on their own under the CCP agreement. The editors condemned the "arbitrary action of the government, taken in haste without giving the bituminous operators a single opportunity to present their case. . . ." The *New York Times* criticized "the sweeping order" for possibly setting prices too low to stimulate production. Since wheat and coal were the "only commodi-

71. *New York Times*, August 11, 12, 20, 1917; Baker, vol. 7, p. 226; Senate Committee on Manufactures, *Shortage of Coal*, pp. 90–91, 148–51. Insull to Wilson, August 17, 1917; Mississippi Council of Defense to Wilson, August 18, 1917; Illinois Governor Frank Lowden to Wilson, August 16, 18, 1917; Wilson to Paul O. Husting, August 21, 1917, Reel 292, Wilson Papers; Fort to Wilson, August 18, 1917, Reel 308, Wilson Papers.

72. The Federal Trade Commission tables on bituminous coal costs used by Wilson in determining prices are in Subject File, U.S. Fuel Administration, Box 126, Garfield Papers; *Coal Age* 12 (August 25, 1917): 321; Wilson to Fort, August 21, 1917, Reel 308, Wilson Papers.

ties" where the price was fixed "by edict rather than by agreement," the *Times* counseled the president that he should feel "no reproach" if they had to be raised.[73]

Conclusions

Long unable to achieve its cooperative ideal by stabilizing either production or prices, the fragmented bituminous industry thus fell under far-reaching wartime controls. By late June, 1917, under the Lane-Peabody committee, volunteers had begun to coordinate transportation and production, but the CCP maximum-price agreement, which affected only noncontract coal, seemed to men of the progressive antitrust tradition to be a conspiracy of coal barons to increase prices. Some coal dealers and coal operators may have been withholding coal from the market, but such actions had little impact on price compared with the exorbitant force of wartime demand. Justly angered by the price increases, consumers attacked the "coal trust."[74]

Without a national organization and national leadership, the industry could not effectively present a favorable image of itself. Moreover, without an established trade organization with political influence, the operators had little clout with the Wilson administration aside from the backing of Franklin K. Lane, a man toward whom Wilson was not overly friendly. Support from Senators Philander C. Knox and Boise Penrose reinforced the industry's image as one of the "big interests." Better-organized industries successfully established working relationships with the Wilson wartime bureaucracy. The highly competitive bituminous industry could not even mobilize political influence to affect significantly the language of the Lever Act, which, in the words of one senator, delegated "greater power over business than

73. *Coal Age* 12 (September 1, 1917); *New York Times*, September 21, October 8, 1917.
74. Susan Armitage, *The Politics of Decontrol of Industry: Britain and the United States*, London School of Economics Research Monograph No. 4 (London: Weidenfeld and Nicholson, 1969), p. 109.

was ever before delegated in a free land in the history of the world—to the President of the United States."[75]

The coal operators, in fact, fell victim to their own disorganization, to their poor public image, and to their own hubris. As a final defense of the CCP prices, *Coal Age* argued obtusely that the industry should not be blamed "for taking a profit when the opportunity came along once in a generation." Senator Francis Newlands noted that the industry had indeed brought controls upon itself. If the operators on the Lane-Peabody committee "had been wise," he said, "if their notion of patriotism had coincided more closely with public opinion," they might have avoided special legislation.[76]

Before American entry into the war, the politics of soft coal hinged on the concept of "the public" versus "the interests," in this case the "coal trust." In these terms the weakness of the industry in the face of the challenge of the progressives who wanted to regulate it casts some doubt on general arguments about the power of large industries in blocking or weakening regulatory legislation. Consumers, particularly representatives of the railroads and the utilities, in the name of "the people" carried the day against the coal men.

The story of the coal industry before the war, however, does fit Gabriel Kolko's generalization about the trend toward industrial rationalization among American industries at the turn of the century.[77] Like many other industrialists, the coal operators tried many ways to employ government to achieve stability. But, given a chance at "industrial self-government" through the CCP, they erred. Wilson intervened, and with the passage of the Lever Act the industry faced the new challenge of trying to influence a government bureau.

75. The creation of the National Coal Association in August, 1917, is treated in chapter 3. Baker to Wilson, August 16, 1917, Box 4, Baker Papers; *Congressional Record*, 65th Cong., 2d sess., 1918, vol. 56, p. 921.

76. *Coal Age* 12 (September 1, 1917); Newlands is quoted in the *New York Times*, July 29, 1917.

77. Gabriel Kolko, *The Triumph of Conservatism* (Glencoe: The Free Press, 1963), chapters 1 and 2.

3

Stabilization under the Fuel Administration

UNDER THE LEVER ACT, President Wilson had the power to regulate the industry as he saw fit. But the government soon found that it needed business expertise. A serious coal shortage during the winter of 1917–1918—brought on in part by the Wilsonians' destruction of the cooperative program of the summer—forced the administration to call coal industrialists into government to facilitate regulation. With industry advice the government ended cross-hauls, helped the industry boost production to the highest levels in its history, and improved car supply while at the same time preventing profiteering by carefully regulating prices. The fragmented industry found during war the stabilization that it had sought in peace. It could not achieve enough power, however, to co-opt or control government.

The Lever Act gave government plenary power over the industry and empowered Wilson not only to fix prices for coal and coke and to license operators, but also to requisition the businesses of those who conducted their firms "in a manner prejudicial to the public interest." He could even force all operators to sell to the government for resale through a pool. The head of distribution under the Fuel Administration told a Senate committee that no other industry in the United States "was controlled, regulated, and directed as specifically and definitely and in such detail as was coal."[1]

1. Lever Act, 40 Stat. 284–86 (1917); James D. A. Morrow's testimony in U.S., Congress, Senate, Committee on Manufactures, *Publication of Production and Prof-*

The law gave Wilson various options. He had used the detailed cost work of the Federal Trade Commission in fixing prices. Commissioner William B. Colver suggested that Wilson use the commission to pool all coal. Instead, Wilson created the Fuel Administration. Two days after he fixed prices, he issued an executive order naming Harry A. Garfield fuel administrator.[2]

By denying the position to William B. Colver, who had expected it, Wilson pleased the operators, but he refused to appoint a coal industrialist. He passed over his own son-in-law, Secretary of the Treasury William Gibbs McAdoo, and—partly on the advice of Herbert Hoover—selected old friend and fellow academician Harry A. Garfield, who was currently working out wheat prices for the Food Administration. Wilson wanted an administrator like Hoover who could view the coal problem from above, putting the needs of a nation at war ahead of the desires of the men of capital.[3]

Son of President James A. Garfield, the fuel administrator brought a background in law and education to the Wilson administration. Young Harry had lived in the White House and had known all the presidents since Ulysses S. Grant. Stunned by his father's assassination, Garfield had searched for a meaningful career. He earned his B.A. at Williams College, his father's alma mater, and returned to St. Paul's, his

---

its in Coal, Hearings on S. 4828, 66th Cong., 3rd sess., 1921 (henceforth cited as *Publication of Production and Profits*), p. 212; William F. Willoughby, *Government Organization in War Time and After* (New York: D. Appleton-Century Co., 1919), p. 296; Paul Garrett, *History of Government Control of Prices* (Washington, D.C.: Government Printing Office, 1920), pp. 160–61; Charles R. Van Hise, *Conservation and Regulation in the United States during the World War* (Washington, D.C.: Government Printing Office, 1917), pp. 54–57; and John Back McMaster, *The United States in the World War*, 2 vols. (New York: D. Appleton & Co., 1929), vol. 1, pp. 374–75.

2. U.S., Fuel Administration (henceforth cited as FA), *Publication No. 1*, August 23, 1917. Memorandum for Dr. Garfield by H. D. Nims, in Records of Harry Garfield, Interoffice Memoranda, RG 67, NA; Garfield to Wilson, February 21, 1917; Wilson to Garfield, February 25, 1917, Records of Bentley W. Warren, RG 67, NA; U.S., Congress, Senate, Committee on Manufactures, *Shortage of Coal, Hearings before the Subcommittee of the Committee on Manufactures Pursuant to S. Res. 163*, 65th Cong., 2d sess., 1918 (henceforth cited as *Shortage of Coal*), pp. 163ff.

3. Herbert Hoover, *The Memoirs of Herbert Hoover*, vol. 1, *Years of Adventure, 1874–1920* (New York: The Macmillan Co., 1952), pp. 243, 253, 262–63; Newton D. Baker, secretary of war, to Wilson, June 27, 1917, Box 4, Baker Papers.

own preparatory school, as a teacher of Roman history and Latin. He earned his law degree from Columbia.

In Cleveland, Ohio, in partnership with his brother James R. Garfield and friend Frederick C. Howe, he built a lucrative law practice. A leader in the Cleveland bar, he organized and became president of the Cleveland Trust Co., became president of the Cleveland Chamber of Commerce in 1898, and led the Cleveland Municipal Association—a progressive group—to victory over "Boss" McKisson. Garfield also taught contract law at Western Reserve University and headed a syndicate which opened the Piney Fork coal mine in southeastern Ohio.

Garfield's interest in politics, his character, and his abilities impressed Woodrow Wilson, then president of Princeton, enough that he offered Garfield the Chair of Politics in 1903. After five years of teaching at Princeton, Garfield assumed the presidency of Williams College, from which he took a leave of absence during the war.[4]

Both visionaries, Wilson and Garfield shared in a search for community through public endeavor. Garfield had read, admired, and recommended to Wilson the collectivist theories of socialist engineering wizard Charles Steinmetz, as presented in his *America and the New Epoch,* which appeared in 1916. Garfield and Steinmetz both thought the war rang the death knell for chaotic individualism in economic and political life. America, they agreed, should attempt to coordinate democracy with some kind of orderly corporate consolidation. Steinmetz even outlined a system of dual government, political and economic, which Garfield would later champion.

Garfield's commitment as fuel administrator to cooperation and coordination in business-government relations provides a key to his later actions. Garfield sought efficiency in

4. Robert D. Cuff, "The Wilsonian Managers" (unpublished manuscript), pp. 3–13; *National Cyclopedia of American Biography* (New York: James T. White and Co., 1954) vol. 33, pp. 154–55; Garfield testimony in *Publication of Production and Profits,* p. 1290; Wilson to Garfield, December 3, 1918, Reel 293, Wilson Papers; Thomas R. Shipp, "Dr. Garfield, Fuel Administrator," *World's Work* 35 (November, 1917): 99–100; Robert D. Cuff, "Harry Garfield, the Fuel Administration, and the Search for a Cooperative Order During World War I," *American Quarterly* 30 (Spring, 1978): 40.

the name of "useful service to mankind." A patrician, called by reformer Frederick C. Howe a man of Roman *virtu*, Garfield was not initially a spokesman for the coal industry, but its mentor in directing it toward orderly and efficient development in the national interest. Relatively overlooked in the histories of the war years, Harry A. Garfield, a bespectacled, slightly balding former professor, carried a great deal of weight with Woodrow Wilson.[5]

Once his work on the wheat prices was completed in early September, Garfield recruited his top staff. At Wilson's suggestion he appointed Leonard A. Snead, a young coal sales manager with a flair for oratory and organization, as the director of apportionment and distribution. Although Snead convinced both Garfield and Wilson to waive civil service requirements and to bring some knowledgeable coal men into the distribution work of the Fuel Administration, Garfield filled his own staff with personal friends and university types. Boyhood companion and fellow Williams alumnus Bentley Warren became legal advisor. Their fathers had served in the same Congresses, and, as Garfield wrote, he and Warren "were as close to each other as men can well be." Another Williams man, lawyer Harry D. Nims, volunteered to help Garfield set up the Washington administration. Although lacking a background in the coal industry, Nims emerged as "Garfield's right-hand man" and later became chief of the legal department.[6]

5. For an expression of Garfield's own commitment to social progress, see FA, *Final Report of the United States Fuel Administrator, 1917–1919* (Washington, D.C.: Government Printing Office, 1921) (henceforth cited as *Final Report*) pp. 23–24, and Appendix IV, pp. 155–56; Baker to Wilson, June 27, 1917, Box 4, Baker Papers; Wilson to Garfield, December 3, 1918, Reel 293, Wilson Papers; Pierpont B. Noyes to Garfield, December 26, 1918, Records of Harry A. Garfield, Interoffice Memoranda and Communications, RG 67, NA; Bernard Baruch, *Baruch: The Public Years* (New York: Pocket Books, 1960; originally published by Holt, Rinehart and Winston, 1952), p. 82; Charles Steinmetz, *America and the New Epoch* (New York: Harper and Brothers, 1916); Josephus Daniels, *The Cabinet Diaries of Josephus Daniels*, ed. David E. Cronon (Lincoln: University of Nebraska Press, 1963), p. 159; Cuff "Wilsonian Managers" and "Harry Garfield," passim.

6. Interview with Leonard A. Snead, February 7, 1974; Garfield to Eugene H. Outerbridge, president, New York City Chamber of Commerce, January 18, 1918, Reel 293, Wilson Papers; Garfield to Josephus Daniels, March 3, 1921, Garfield Papers; FA, *Final Report*, p. 9; Garfield testimony in *Shortage of Coal*, p. 9; *Coal Age* 12 (November 24, 1917): 898.

Experts from government and universities filled other advisory posts. On loan from the Geological Survey, C. E. Lesher and six associates helped Leonard Snead untangle supply problems. With a staff of six hundred to handle statistics on supply and prices, Lesher made painstaking tabulations that are still the best data on prices and production for the war years. Samuel A. Taylor, a mining engineer and dean of Allegheny College, became technical advisor. As his private secretary, Garfield chose Van H. Manning, chief of the Bureau of Mines.[7]

Fearing appointment of a "radical" as fuel administrator, operators were relieved at Garfield's appointment. Even though he initially spoke of even further reducing consumer prices, *Coal Age* hoped that "with the proper technical guidance, Dr. Garfield will be equal to the tremendous problem which confronts him." Still smarting from Wilson's unilateral two-dollar price, which had driven spot coal from the market and slowed production, *Coal Age* complained, "A pall of uncertainty and foreboding hangs dark over the land." Continuing the Wilson prices, the journal warned, would "inevitably drive all but the larger and more efficient mines upon the rocks of bankruptcy."[8]

Garfield assuaged industry fears somewhat by consulting with Francis S. Peabody, former chairman of the Committee on Coal Production. Still, *Coal Age* bemoaned the "serious lack of experienced coal men on the staff of the fuel administrator." The journal detected throughout the Wilson administration "distrust of the Nation's most successful business men." In December, Garfield's early disputes with operators—he compared some to the "robber barons of the Rhine"—led him to propose nationalization of the mines.[9]

Both the Lever Act and Garfield insured the primacy of government in the early stages of wartime regulation of coal. Wilson still planned for an eventual government pooling of

7. Garfield to Daniels, March 3, 1921, Box 86, Garfield Papers; FA, *Final Report*, p. 26.

8. *Coal Age* 12 (September 1, 1917): 369, 376.

9. Ibid. (September 8, 1917): 412; 13 (January 5, 1918): 26; (February 2, 1918): 253; *New York Times*, September 16, November 25, December 27, 31, 1917; Willoughby, p. 297; interview with Leonard A. Snead, February 7, 1974.

all coal not under contract, as proposed by William Colver. By fixing "tentative" prices himself, the president had acted under a clause in the statute which expressly abrogated existing contracts. Only the "tentative" nature of the prices allowed the industrialists to believe that their contracts, covering some three-fourths of the tonnage, might be maintained. To early December, 1917, the coal industrialists had failed to exercise any significant influence over Wilson's prices, the Lever Act, or Garfield.[10]

But weather and the government's need for expertise would draw the industrialists and the Fuel Administration together. One of the worst winters in the nation's history forced Garfield to turn for help to the industry Wilson had chastised. At the same time, the industry realized that central direction under government controls not only was in the national interest but also meant better marketing and improved production. War reversed the economic position of the industry. Called the "conflict of the smokestacks" by Secretary Baker, war meant that the nation and the allies would consume as much coal as the industry could mine and the railroads could transport.[11] Thus, the industry and the government both sought improved use of railroad cars, coordinated distribution, and an end to cross-hauling. A new association soon provided the government with a pool of talent.

The National Coal Association

Movement to create a national association of coal operators developed during the summer of 1917 as the meetings of the Committee on Coal Production were being held in Washington. Once the CCP established the Lane-Peabody minimums, a committee of trade association secretaries chaired by C. P. White of Cleveland, including C. E. Lesher of the Geological Survey as secretary, decided to create a national coal association as a clearinghouse to gather statis-

10. Robert A. Taft to Garfield, August 30, 1917; Memorandum for Dr. Garfield by H. D. Nims, Records of Harry Garfield, Interoffice Memoranda, RG 67, NA; Lever Act, 40 Stat. 284 (1917); *Coal Age* 12 (September 15, 1917): 452.

11. Baker quoted in Mark Sullivan, *Our Times: The United States, 1900–1925*, vol. 5, *Over Here* (New York: Charles Scribner's Sons, 1933), p. 372, fn. 1.

tics, make "the price-fixing plan readily operable," publicize information on the industry, and coordinate industry activities with the bureaus and departments of the Wilson war administration. Although initially fearful of antitrust laws, the founders argued that promoting "a harmonious feeling" among member associations was "entirely within the law."[12]

During July and early August—as the CCP committees attempted to step up production but before Wilson had quashed the three-dollar price—the industry leaders organized the National Coal Association (NCA). Representatives of thirty-one of the industry's forty-three coal trade associations hammered out a declaration of purposes. The industry needed a national association, they believed, for conservation of coal, cooperation with governmental officials, collection of data, enforcement of state and federal laws, institution of improved cost accounting, establishment of information bureaus, maintenance of adequate car supply from the railroads, and the general welfare of the coal industry. "No part of the machinery of this association," declared the founders, "will be permitted to be used to establish, regulate, maintain or control prices for the sale or to divide territory, to regulate, diminish, or control the production of coal, or limit or control competition. . . ."[13]

By the last week in August the NCA had elected its directors and claimed formal membership of twenty associations and 610 operating companies, primarily from the Central Competitive Field, together producing about 30 percent of the nation's coal. Organization-minded executives from the older mining areas, representing the larger coal firms, grasped an opportunity brought by the crisis of war. The directors included Henry N. Taylor of Central Coal and Coke, active in the CCP, and Rembrandt Peale of Peale, Peacock,

12. Interview with Leonard A. Snead, February 7, 1974; copy of a speech given by Mr. Snead on the fiftieth anniversary of the founding of the NCA, Snead Papers, privately held; E. Pendleton Herring, *Group Representation before Congress* (Baltimore: Brookings Institution, 1929), p. 103; *Coal Age* 12 (July 7, 1917): 33.

13. *New York Times*, August 22, 1917; David B. Truman, *The Governmental Process* (New York: Alfred A. Knopf, 1962), p. 176; James D. A. Morrow's testimony and exhibits in *Publication of Production and Profits*, pp. 119–23, 2060–67; *Coal Age* 11 (June 30, 1917): 1124; 12 (August 8, 1917): 202–203.

and Kerr. Pittsburgh Coal's W. K. Field became the president, and Jere H. Wheelwright of Consolidation Coal became vice-president. But all were really second to the dynamic and aggressive James D. A. Morrow, who had most recently been working for Edward N. Hurley in the FTC. An Ohioan with a degree from Ohio Wesleyan, Morrow entered the coal business in Pittsburgh, rose to be the commissioner of the Pittsburgh Coal Producers Association, and then became assistant secretary of the FTC, reporting to Hurley. He backed Hurley's plan to improve efficiency in coal company accounting and helped initiate the FTC's investigation of coal production, but like Hurley, he opposed Colver's pooling recommendations. Morrow's sharp features, rimless spectacles, largish ears, and dark hair neatly parted right down the middle of his head gave him an austere appearance. A forceful man then in his thirties, and an exceptional organizer, Morrow led the coal industry during both the war and the New Deal.[14]

By mid-winter the NCA more than doubled its membership and established offices in Washington. It printed a *Daily Digest* of Fuel Administration rulings and announcements, FTC reports, ICC bulletins, and daily advice on car supply and shortages. Its releases argued that the shortage of railroad cars caused what consumers saw as a "coal shortage." The NCA spurred a drive to improve cost accounting and printed a manual to help operators complete FTC forms on production and costs. The NCA's traffic department pressured the war bureaucrats for better car distribution.[15]

A divided, unsure NCA sought to define its relationship with the Fuel Administration. Some members wished to fight Wilson's prices in the courts. Others, like James D. Francis, president of the powerful Island Creek Coal Co., believed that reasonable profits could be earned even with the two-dollar maximum price. When some Michigan operators refused to ship coal at government prices, Garfield threatened

14. Garfield memorandum, August 29, 1919, Special Correspondence, Box 93, Garfield Papers; *Coal Age* 12 (August 25, 1917): 321; (September 29, 1917): 537; 13 (February 9, 1918): 290.
15. Arthur E. Suffern, *The Coal Miners' Struggle for Industrial Status* (New York: The Macmillan Co., 1926), p. 179; *Coal Age* 13 (January 19, 1918): 114.

to take over their mines, but patriotic pressures by Leonard Snead brought some tentative operator acquiescence. Once the government helped mediate a wage dispute and granted a price increase to cover costs, business-government relations began to improve—but only for the moment.[16]

The Winter Crisis

Garfield had unwisely announced that coal production would be sufficient for the war needs of the winter of 1917–1918. But coal operators knew that the lack of railroad coal cars could well slow production. Bernard Baruch warned Wilson of an unavoidable shortage. Under the powers of the Preferential Shipments Act, passed the same day as the Lever Act, Wilson appointed Judge Robert S. Lovett, a Texan with long experience in railroading, as priority commissioner. In late August, Lovett directed the railroads to give preference over other commodities to bituminous coal traveling to Great Lakes ports. But November coal production improved, particularly in the Midwest, and the order was revoked.[17] To meet New England coal shortages, Garfield ordered all mines with New England contracts to deliver maximum monthly requirements to Hampton Roads and Baltimore ports for water shipment to New England. Under NCA pressure, Garfield finally got the chairman of the General Operating Committee of the eastern railroads to give preferential distribution of open-top cars to the mines.[18]

16. *New York Times*, August 30, November 25, 1917; H. D. Nims memorandum for Garfield, October 22, 1917, Records of Harry A. Garfield, Interoffice Memoranda, RG 67, NA; *Shortage of Coal*, pp. 31, 184; Van Hise, pp. 150–51; Leonard Snead recalled winning one operator's cooperation by comparing the coal man's potential loss with that of some young recruits then marching past Fuel Administration headquarters. Snead interview, February 7, 1974. The Washington Agreement is treated more fully in a later section of this chapter.

17. Willoughby, pp. 92–98; Wilson to Garfield, September 5, 1917, with attached memorandum from Bernard Baruch, Box 91, Garfield Papers; Wilson to Baker, August 17, 1917, Box 4, Baker Papers; Wilson to Garfield, August 27, 1917, Reel 292, Wilson Papers; Frederic L. Paxson, *American Democracy and the World War*, vol. 1, *America at War* (Boston: Houghton-Mifflin Co., 1939), p. 255; *New York Times*, October 11, 1917; Robert D. Cuff, *The War Industries Board: Business-Government Relations during World War I* (Baltimore: Johns Hopkins University Press, 1973), pp. 119, 192–94; *Coal Age* 12 (December 1, 1917): 934.

18. National Coal Association Resolution, October 18, 1917, Records of Harry A. Garfield, General Correspondence, RG 67, NA; *Coal Age* 12 (December 8, 1917): 976.

Car shortage, however, worsened. Coal famines developed. The *New Republic* called the Wilson administration's failure to coordinate coal priority one of the "most astonishing and scandalous instances" of bureaucratic failure in the war. Garfield sought to have Lovett issue a new nationwide priority order favoring coal and open-top cars for mines, but Food Administrator Herbert Hoover protested giving preference to coal "or anything else" over food.

Inability to move food, Hoover feared, might break prices his administration had set. Shortages of cars for food, he announced, were "a matter of most extreme anxiety. . . ." On December 8, 1917, Lovett finally issued a major railroad priority order. Effective December 12, all railroads were to follow the following priorities: first, steam coals for current railroad operation; second, livestock, perishable freight, food, and supplies; third, military supplies consigned directly to the government, army, navy, U.S. Shipping Board, or the Allies; fourth, coal for by-product plants; and fifth, coal for current use in hospitals, schools, and other public institutions or for domestic consumers, blast furnaces, foundries, or manufacturers engaged in war work.[19]

Lovett's order, said James D. A. Morrow, was "a gold brick." By leaving other commodities ahead of coal (except steam coal for railroad engines), the priority order freed no cars for the coal industry. No coal cars mean no coal—and no manufacturing. In an appeal to Lovett to reconsider his order, Garfield argued that the coal shortage stemmed from railroad congestion at terminals, which prevented unloading of coal already on trains. Only by unloading coal and rushing available open-top cars back to the mines could the congestion be eliminated. "The needs of our householders," warned Garfield, "are leading people to riotous outbreak in many quarters, nor is this to be wondered at when it is recalled that thousands of loaded cars are standing on our tracks while coal-bins are empty."[20]

19. *New Republic* 13 (December 8, 1917): 141–42; Daniels, *Cabinet Diaries*, January 14, 1918, p. 267; *New York Times*, December 1, 4, 1917; April 11, May 14, 1918; Willoughby, p. 310; Garfield to Wilson, November 26, 1917, Box 92, Garfield Papers; *Coal Age* 12 (December 8, 1917): 977; (December 12, 1917): 1081.

20. *Coal Age* 13 (January 5, 1918): 10–12; Garfield to Lovett, December 24, 1917, Box 92, Garfield Papers.

To help insure equitable local distribution of coal, Garfield had appointed leading citizens, generally politicans or civic leaders not connected with the industry, as state fuel administrators. But where widespread shortage developed, people panicked. In Ohio, where priority shipment of coal to Great Lakes ports was moving past cities without coal for home heating, citizens tore up railroad tracks to prevent shipment of coal out of their state. The Ohio fuel administrators found nearly one million tons of bituminous "hidden" near Cleveland in unauthorized storage areas. Garfield quickly ended shipments to the firms which had accumulated surpluses. In November the mayor of Willoughby, Ohio, held up a freight train laden with coal and arrested the engineers and conductors. The "law of humanity," he argued, "took precedence over property rights." In December, James Cox, Ohio's Democratic governor, ordered that "solid train loads of coal" be assembled for his state's needs. The NCA reminded the public that a shortage of cars caused the Ohio crisis.[21]

In December, temperatures plunged to fourteen below zero in Boston and elsewhere in the East. Railroad snarls and frozen harbors compounded transportation problems. Dependent on water transportation for its coal, New England was hit hard. Ice closed the harbor in Stamford, Connecticut, and school officials extended the Christmas vacation for Stamford's public schools by three weeks. School leaders elsewhere discussed terminating the entire school year at Christmas. The cold weather froze coal in dealers' lots. When would-be purchasers in Brooklyn failed to get coal, they rioted, hurling bricks, stones, and coal scuttles at merchants. Several hundred New York men, women, and children stormed coal wagons on Manhattan's East Side. Coal pushed ahead of food in the public mind as the single most important commodity in the struggle to defeat the Kaiser.[22]

21. *New York Times*, October 8, 18, 23, November 14, December 13, 14, 17, 1917; *Congressional Record*, 65th Cong., 2d sess., 1918, vol. 56, part 1, p. 914; *Shortage of Coal*, pp. 69–70; Van Hise, p. 158.

22. Final Report of Thomas W. Russell, in FA, *Report of the Administrative Division* (Washington, D.C.: Government Printing Office, 1920), p. 43; *New York Times*, December 13, 1917; January 2, 1918; "Situation Report," December 31, 1917, Reel 293, Wilson Papers; *Coal Age* 12 (December 29, 1917): 1091; Frederick Palmer, *Newton D. Baker: America at War*, 2 vols. (New York: Dodd, Mead & Co., 1931) vol. 2, p. 31.

Following one of the coldest days in the history of New York, the *New York Times* reported on January 1, 1918, that New Yorkers generally had but one day's supply of coal left. The day before, New York State Fuel Administrator Albert Wiggin decreed that all outdoor electric lights in New York State—save those required for public safety—be extinguished every night except Saturday. The New York Edison Company's inspector threatened personally to cut the wire of those who violated the order. Fearful that the cold weather would end river transportation of coal, Wiggin telegraphed Garfield to open the Pennsylvania Railroad passenger tunnels to allow coal to move to Queens, Brooklyn, and Long Island. Within hours, Samuel Rea, the railroad's president, had coal moving instead of people. A resourceful New Jersey mayor brought coal into the city in his street cleaning trucks.[23]

"Suffering, sickness, and possible deaths imminent and inevitable," the Worchester, Massachusetts, fuel committee wired the president, "unless change in system of supplying coal to our district is put into effect." In Detroit, one civic leader carried coal to the needy in his limousine. Motor City doctors reported that patients with serious illnesses lacked fuel to warm their homes in the sub-zero weather. When coal ran out, some Detroiters panicked and abandoned their homes. One who remained in an unheated house froze to death.[24]

The cold compounded the railroad problem. Industrial and passenger use of the railroads had mushroomed with war demand, putting serious strain on an underfinanced and undermaintained industry. Railroad workers had gone to better-paying jobs in war industries. In December the rail congestion, the cold, and the priority snarls paralyzed the nation's rail system. On December 26, 1917, Wilson abruptly placed

23. Sullivan, vol. 5, p. 636; *New York Times*, January 1, 2, 6, 1918; *Edison Weekly,* January 9, 1918, p. 3.
24. Worcester, Massachusetts, Fuel Committee to Wilson, December 30, 1917, Reel 293, Wilson Papers; Homer Warren to Thomas J. Ansatell, December 14, 1917; Joseph H. Hathaway, Highland Park, Michigan, health officer to Highland Park Chief of Police Seymore, December 14, 1917, Reel 292, Wilson Papers; *Literary Digest,* January 5, 1918, p. 18.

control of the industry in the hands of Secretary of the Treasury Williams Gibbs McAdoo, whom he made railroad administrator. Garfield and McAdoo both supported the president's move.[25]

After conferring with Garfield, various railroad presidents, Secretary of War Newton D. Baker, Secretary of the Navy Josephus Daniels, and the president, McAdoo made coal number one in priority. In a closed session, the NCA executive directors supported McAdoo. But altering priorities did not immediately unsnarl the congestion at the eastern terminals, where ships that were headed for war ports awaited coal. On January 5, winds of fifty-five miles an hour brought the first of two massive snow storms to the Midwest. Extreme cold froze switches, ice clogged harbors, and whole trains of coal stood outside terminals, either frozen or blocked from unloading by the congestion of other freight.[26]

Hundreds of ships—many carrying munitions and other supplies destined for the Allies—lay in harbors waiting for bunker coal. Garfield dispatched P. B. Noyes, his chief of conservation, to Boston and New York to investigate. Noyes's long-distance call of January 16 moved Garfield toward immediate and drastic action. Fuel Administration officials discussed establishing "fuelless Mondays." In order to relieve congestion at the ports, Garfield planned to order that no manufacturer or manufacturing plant east of the Mississippi be allowed to burn coal between January 18 and 22 and on every Monday thereafter until March 25. Only by stopping the flow of manufactured goods, he reasoned, could the rail-

25. McAdoo to Wilson, December 6, 15, 1917, Correspondence, Box 523, McAdoo Papers; *Coal Age* 12 (November 24, 1917): 893; (December 22, 1917): 1058; K. Austin Kerr, *American Railroad Politics, 1914–1920: Rates, Wages and Efficiency* (Pittsburgh: University of Pittsburgh Press, 1968), pp. 27–28, 48, 64–72; K. Austin Kerr, "Decision for Federal Control: Wilson, McAdoo, and the Railroads, 1917," *Journal of American History* 54 (December, 1967): 550–60; Garfield to Wilson, November 26, 1917, with "Memorandum Concerning the Production of Coal," Box 92, Garfield Papers; I. L. Sharfman, *The American Railroad Problem* (New York: The Century Co., 1921), pp. 94–99; Walker D. Hines, *War History of American Railroads* (New Haven: Yale University Press, 1928), pp. 12–21; Albro Martin, *Enterprise Denied: Origins of the Decline of American Railroads, 1897–1917* (New York: Columbia University Press, 1971).

26. Ray Stannard Baker, *Woodrow Wilson: Life and Letters*, 8 vols. (Garden City, N.Y.: Doubleday, Page, 1927, and Doubleday, Doran, 1927–1939), vol. 7, p. 437; *New York Times*, December 28, 1917; January 1, 1918.

roads unload cargoes and free the rails so that coal could be moved into the ports to fuel the waiting ships.[27]

On January 16 Garfield met with Secretary of War Baker and told him of the great industrial shutdown he planned to impose. Startled, Baker told him that it "would loose a whirlwind" but agreed to arrange a secret meeting with Wilson later that afternoon. At the White House, Baker, Secretary of the Navy Daniels, and the president heard Garfield out. They implemented minor changes in the order to protect fuel supplies for the navy and other phases of the war program. Garfield left with the approval of those present. Wilson suggested that the fuel administrator clear the measure with McAdoo. He did and rushed back to the Fuel Administration offices to announce his decision before the press got the story. Because his legal department could not get a copy of the order ready for release, Garfield issued an abstract that evening and hurried to keep a dinner appointment with the Hoovers.[28]

On the seventeenth, Congress and the executive branch locked horns. Wilson's congressional opponents tried to block the order. Garfield met unsuccessfully with a group of senators that afternoon to remove their doubts. One senator remarked, "I rather think that a million German soldiers turned loose in this country would not create the havoc that this order will create if enforced." Senator Gilbert Hitchcock of Nebraska, Wilson's most outspoken foe, got Senate acceptance of a resolution delaying implementation of the order for five days by a vote of 57–19. The House passed a resolution expressing its regret over Garfield's "summary

27. FA, *Final Report*, pp. 32–33; FA, *Publication No. 17* (Washington, D.C.: Government Printing Office, 1918); *Coal Age* 13 (February 9, 1918): 273; *New York Times*, January 16, 18, 1918; Garfield to Daniels, March 3, 1921, Box 86, Garfield Papers; *Congressional Record*, 65th Cong., 2d sess., 1918, vol. 56, part 1, pp. 913–14; A. B. Daniels, president, American Paper and Pulp Association, to Wilson, January 16, 1918, Reel 293, Wilson Papers.

28. Baker, vol. 7, p. 475; Garfield to the vice-president of the Senate, January 18, 1918, Box 93, Garfield Papers; Newton Baker to Garfield, March 31, 1922, Box 92, Garfield Papers; Garfield to Daniels, March 3, 1917, *Congressional Record*, 65th Cong., 2d sess., 1918, vol. 56, Part 1, pp. 912–36, 997–1004; Joseph P. Tumulty, *Woodrow Wilson as I Knew Him* (Garden City, N.Y.: Doubleday, Page and Doran, 1921), pp. 360–63; *New York Times* January 18, 1918; FA, *Final Report*, pp. 32–33.

action" and asked Wilson to reconsider. Aware that the Senate was about to act, Garfield signed the formal order shortly before six o'clock. The Senate messenger bearing the resolution reached him only minutes afterward.[29]

Some newspapers tried to look on the positive side of the order. One wrote that it would not have been issued "unless it were felt to be absolutely necessary." But most of the press agreed with Congress. The *New York Times* called it "unnecessarily harmful and destructive." The *New York World,* normally sympathetic to the administration, jibed that at a dollar a year, Garfield was apparently worth it. They called the order the "greatest disaster that has befallen the United States in this war" and believed it "worthy of a Bolshevik Government." The *Boston Globe* wrote, "Picture the Germans charging across No Man's Land at you with bayonets fixed and blood in their eyes, and imagine your feelings if you were behind a machine gun and were ordered not to fire a shot for five days—to save ammunition." The "Fuel Administration," said the New York *Tribune,* "has lost its head." In Providence, Rhode Island, the *Journal* blamed the act on "panic-stricken incompetency." The *New Republic* believed Garfield to have "infirm judgment."[30]

Once the public was aware that Wilson backed Garfield, opposition weakened, and the five-day ban on manufacturing and Monday ban on all commercial fuel use went into effect on the eighteenth. By then Garfield had permitted about one thousand exemptions, but he had angered thousands more with denials. In the New York area, one-half million industrial workers stayed home for the five days, and hundreds of thousands more did not work on "heatless" Monday, when the stores were closed. The bullish man-

29. *Congressional Record,* 65th Cong., 2d sess., 1918, vol. 56, part 1, pp. 912–36, 997–1004; Republicans voted twenty-eight to three in favor, Democrats twenty-two to sixteen in favor.

30. The newspaper opinion is summarized in *Literary Digest,* January 26, 1918, pp. 6–8; *New York Times, New York World, New York Tribune,* January 18, 1918; *New Republic* 13 (January 26, 1918): 360–61. Frank Cobb, editor of the *World* and a long-time friend of the president, told him: "The whole of New York, both rich and poor, is seething in protest. The whole psychology of it is disastrous. It is demoralizing the whole country" (Cobb to Wilson, dictation of a telephone message, January 17, 1918, Reel 292, Wilson Papers).

agers of the New York Stock Exchange determined to stay open without heat; traders wore overcoats and heavy sweaters on the floor; office assistants at the exchange worked by candlelight. Barflies kept warm in unheated bars by their own techniques. At night, New York's lightless streets seemed to one observer like the Sahara between moons.[31]

Although many celebrated the "Garfieldays" by attending the extra matinee stage shows and movies scheduled on Mondays, a million and a half individuals lost wages. Garfield asked that employers maintain the normal weekly wages. Most refused. One estimate made for the Baltimore and Ohio Railroad held that New Yorkers lost over $350 million in wages, Chicagoans $313 million, Philadelphians $138 million. This calculation set production losses at $1 billion for New York, $381 million for Chicago, and $333 million for Philadelphia. The totals for the eighteen major cities came to $4.3 billion. Estimations of the losses to New York by the New York Board of Trade were in the same range as those of the B&O.[32]

A leading coal journal estimated that the shutdown saved 3,456,000 tons of coal. Measured against the lost wages and production in the eighteen major cities, the cost per ton was $1,256.94. Even if the total loss in wages and production is scaled down to $1 billion, the cost per ton, as critics pointed out, was $289.35. Garfield shot back: "To estimate the value or harm of the order by the dollars' worth of merchandise unmanufactured and unsold and the number of men thrown out of employment temporarily is to fail to understand the crisis."[33]

31. *New York Times*, January 18, 1918. Cuff, *War Industries Board*, pp. 136–37; Paxson, pp. 215–16; final report of Fuel Administrator S. P. Kennedy in FA, *Report of the Administrative Division* (Washington, D.C.: Government Printing Office, 1920), pp. 13–14; *Railway Age* 64 (January 25, 1918): 198; *New York Evening Post*, January 18, 1918; *New York American, New York World, New York Herald, New York Times*, January 22, 1918.

32. *New York World*, January 19–22, 1918; *Economic World* 15 (February 23, 1918): 267; *New York Times*, January 19, 1918.

33. The *Black Diamond*, quoted in *Literary Digest*, March 2, 1918, p. 18. Senator Smith of Michigan estimated that the six tons of coal saved in his home city of Grand Rapids cost twenty-two thousand dollars (*Coal Age* 13 [February, 1918]: 293). The *New York Times* (February 17, 1918) estimated that each one dollar's worth of coal saved cost nine dollars in wages and twenty-three dollars in production. FA, *Final Report*, p. 33.

The order, he protested, aimed not at conserving coal but at opening the rail connections to the port cities to bunker the ships to carry "to the Allies the food and war supplies they vitally need." He claimed that his order had facilitated the movement of 480 ships, including 369 carrying munitions and other war supplies. However, neither Garfield nor his spokesman George Creel, who conducted a propaganda drive on Garfield's behalf, ever supplied lists of ships and sailings. Normally New York longshoremen coaled an average of 12 ships per day. Over the five-day shutdown they pushed the average to 15.4 per day, a notable but not spectacular increase. Eastern railroad tie-ups eased somewhat following the order, but over the long weekend the railroad congestion did not perceptibly untangle.[34]

Garfield's champions stigmatized his detractors as unpatriotic, but his critics were correct: the shutdown order was an unnecessarily radical solution to the problem. For the results accomplished, the costs were overwhelmingly high. *Coal Age* wrote that the extensive railroad embargo on materials other than coal would have broken the railroad congestion. Although McAdoo had moved coal to top priority, he apparently was unwilling to order an embargo on other commodities. Garfield never criticized McAdoo directly, but wrote later that many in the administration saw the approaching crisis but "hesitated to assume responsibility." Director General McAdoo had been in office barely three weeks.[35]

But to focus on the excesses of the order itself is to shift attention from the administration's actions of the late summer. When Wilson ended the price-stabilizing efforts of the Committee on Coal Production, orders and production fell off. The administration's attack disgraced the Committee on Coal Production, which had marshaled the support of the coal community to greater productivity and better distribu-

34. FA, *Final Report*, p. 33; "The Case for Mr. Garfield," *Independent* 93 (March 19, 1918): 408–409. The *New York Times*, January 23, 1918, gave a complete table of the number of ships coaled, cars on the railroads above normal, and tons of coal at New York tidewater ports from January 13 to January 22; *Railway Age* 64 (January 25, 1918): 200.

35. "The Case for Mr. Garfield," pp. 408–409; *New York Times*, March 2, 1918; *Coal Age* 13 (February 9, 1918): 273; FA, *Final Report*, p. 32. Garfield himself had been in charge of the Fuel Administration only since mid-September.

tion. Because the coal men had been encouraged to agree on prices by Secretary Lane, they felt they had been misled and abused. The CCP collapsed. Then Wilson imposed the two-dollar limit and priced out the marginal producers. Garfield later rectified this error by allowing increases for numerous fields, but the initial damages had been done. The nation began the winter of 1917–1918 without its coal stockpiles. The biting cold weather and the rail congestion exacerbated a bad situation, of course, but the administration's attack on the CCP prices and the inability of the government and the railroad men to coordinate transportation priorities, in particular return of coal cars to the mines, caused the shortage in the first place.

The Zone Plan

The January debacle and the ensuing congressional debate about the lack of business talent within the Fuel Administration forced the administration to consider seeking coal industry expertise. In an attack which must have stung both Wilson and Garfield, Senator Henry Cabot Lodge asked rhetorically, "Who is the fuel controller. . .? A college president who never saw a coal mine, probably, in his life and knows no more about coal than when he sees it burn in the stove." In the fall, the Railway War Board and Francis S. Peabody had proposed a zoning plan for coal distribution that they felt could cut cross-hauls and increase production by 20 percent. Railroad Administration (RA) agents discussed the scheme with the Fuel Administration, which had been working on its own plan. Wilson backed a proposal which restricted consumers to coals produced in their own regions. In late January, Garfield and McAdoo met to implement coal zoning.[36]

They divided the country into eleven producing districts and twenty consuming regions. Once the Fuel Administration established how much coal each consuming region re-

36. *Congressional Record*, 65th Cong., 2d sess., 1918, vol. 56, part 2, p. 1091; Wilson to Garfield, January 21, 1918; Garfield to Wilson , January 22, 1918, Box 92, Garfield Papers; Peabody testimony in *Shortage of Coal*, pp. 578–79; *New York Times*, January 15, 24, 1918.

quired, it determined from which producing districts the coal could be most easily and cheaply moved. In essence, the plan required the major consuming interests in the Midwest to purchase coal from their region instead of Pennsylvania, West Virginia, or eastern Kentucky. On nomination from the operators' trade associations in the producing districts, Garfield appointed district representatives—generally coal association secretaries—who would decide on emergency needs and exceptions. The plan, Garfield thought, would eliminate millions of miles of cross-hauling, end congestion in the Appalachian gateways to the West, and also decentralize decision making for the Fuel Administration. *Coal Age* called it "the most important step ever taken in the affairs of the soft-coal industry."[37]

Garfield announced on January 25 that the proposal would take effect on April 1 at the beginning of the new contract year and appointed the NCA's general secretary, J. D. A. Morrow, as head of a restructured Distribution Division. Morrow brought experienced coal men—usually trade association secretaries—into the Fuel Administration as district administrators. To end the numerous controversies that arose where producers lay close to a dividing line that separated them from their traditional markets, Morrow's staff redesigned zones. Coal association agents helped operators establish relationships with new retailers and consumers within their zones. Despite the expected troubles in such a massive restructuring of markets, the plan, *Coal Age* reported in late April, was "being well received by operators." The Railroad Administration cooperated by policing unauthorized shipment of coal across zone lines.[38]

37. Memoranda, October 14, 1917, on state fuel administrators, Reel 293, Wilson Papers; J. D. A. Morrow to Garfield, January 22, 1918, Records of Harry Garfield, General Correspondence, RG 67, NA; *Coal Age* 13 (January 26, 1918): 216; (February 2, 1918): 250–51; (March 2, 1918): 422; FA, *Report of the Distribution Division* (Washington, D.C.: Government Printing Office, 1919), pp. 5–10, 25–27.

38. *New York Times,* January 26, 1918; March 22, 1918; Morrow to Garfield, January 9, 1919, Records of Harry A. Garfield, General Correspondence, RG 67, NA; U.S., Railroad Administration, *Annual Report of W. G. McAdoo, Director of Railroads: Operation* (Washington, D.C.: Government Printing Office, 1919), pp. 35–36, in Box 557, McAdoo Papers; Evans Wollen's report in FA, *Report of the Administrative Division*, p. 99; FA, *Report of the Distribution Division*, pp. 23–31; *Coal Age* 13 (April 20, 1918): 756; (May 11, 1918): 878ff.; Garrett, p. 163.

To correctly apportion intradistrict allocations, the Priorities Committee of the WIB, in negotiation with the Distribution Division of the Fuel Administration, developed WIB Preference List No. 1, issued on April 6, 1918. Delays in determining the significance to war requirements of specific plants delayed the Fuel Administration's direct use of the lists until June. But after that, the Distribution Division and the WIB's Priorities Committee cooperated to produce a workable system. Morrow's district administrators in the Fuel Administration represented producers; the state fuel administrators represented consumers. Together they apportioned the nation's coal.[39]

The zone plan restored the cooperation between government and industry which had broken down under the Committee on Coal Production. Because the 1918 Fuel Administration appropriation could not cover the cost of the new distribution program, Morrow convinced the operators "to let us use their trade associations as distributing agencies for the Fuel Administration, utilizing their information, personnel, offices, etc., with the operators paying all the expenses. . . ." He estimated the government saved approximately $1.5 million. Some attacked this cozy bargain as a clear conflict of interest. But investigations exonerated the coal men. As former trade association secretaries, the district representatives may have been somewhat partial to the leading firms in their districts, but keeping their positions had always required that they be impartial. The Fuel Administration scrutinized their decisions every week.[40]

William Hard, an investigative reporter for the *New Republic* who uncovered the $1.5 million contribution from the industry, praised what he found. Although the *New Republic* had earlier harped on the inefficiency of wartime government bureaucracy, Hard wrote in November, 1918,

39. FA, *Final Report*, p. 30; FA, *Report of the Distribution Division*, pp. 28–35; William Hard, "Socialistic Coal," *New Republic* 17 (November 16, 1918): 64–66.
40. Morrow openly reported the assistance of trade association secretaries in the distribution of coal to William Hard of the *New Republic*. Garfield asserted that Fuel Administration officials controlled all "matters relating to price or policy of distribution." William Hard to Morrow, September 5, 1918; Morrow to Hard, September 13, 1918; Tumulty to Wilson, September 23, 1918; Garfield to Senator Thomas J. Walsh, September 25, 1918, Reel 293, Wilson Papers.

that "the coal industry moves toward becoming a self-conscious unit, a capitalistic Guild, discharging its responsibility to other industries through its representatives who meet the State and who cooperate with the State." Dr. Garfield, Hard wrote, "has not imposed a bureaucracy" on the industry, but has given it "a new dignity by imposing certain definite collective duties on it, in the name of the consumers, and by them summoning it to devise and to operate the methods by which those duties shall be done. He has developed the unit of the producers equally with the unit of the consumers." The distribution program, Hard concluded, "is a triumph of organized units over unorganized individualism."[41] Earlier both operators and consumers had abused Garfield. Now he was cheered for bringing businessmen and bureaucrats together to make the distribution of coal rational and efficient.

Garfield was pleased. He claimed that zoning Pocahontas coal out of Chicago saved an estimated 11.4 million car miles and that another 2.5 million car miles were eliminated by denying Kanawha coal to Wisconsin. Zoning, he wrote, saved almost 160 million car miles, and thus conserved the coal necessary to fire the engines over those miles. The Fuel Administration opened bottlenecks in the Appalachian gateways, allowing 9 percent more coal to move through the mountains, and provided coal to New England, where supplies of coal had been drastically short during the winter of 1917–18. Shorter railroad hauls meant more cars and production. Once the zone system became fully operational during the summer and fall of 1918, weekly coal production reached the highest levels in the industry's history.[42]

Previously facing a buyer's market and unable to reach any cooperative marketing arrangements, the operators could now sell all their output at prices nearly double those

41. Hard, pp. 64–66; on William Hard and progressivism, see Paul W. Glad, "Progressives and the Business Culture of the 1920s," *Journal of American History* 53 (June, 1966): 75–89.

42. A. G. Gutheim, "The Transportation Problem in the Bituminous Coal Industry," *American Economic Review* 11 (March, 1921): 99; FA, *Report of the Distribution Division*, pp. 25–27; Kerr, *Railroad Politics*, p. 82; Willoughby, pp. 308–309; *Coal Age* 13 (June 22, 1918): 1116; 14 (November 21, 1918): 949.

of a few years before. They agreed to the government's dramatic restructuring of their markets with minor misgivings.

The zone plan enhanced Garfield's standing with the operators. He soon won their praise. Although the zone concept included some pooling of railroad cars and improved the car-supply problem of the operators, John Skelton Williams, McAdoo's assistant in charge of finance and purchases, continued to use the traditional railroad man's bait of offering more cars to those coal operators who would submit lower bids on locomotive fuel. Williams argued that the lower prices to the railroads would save the government between one hundred and two hundred million dollars, a figure later reduced in the face of coal industry data. Garfield went to Wilson to argue that unequal car supply shut down high-cost mines, threw miners out of work, and was a "fundamental injustice." He sent a message to the International Railway Fuel Association meeting in Chicago, arguing that "changing times" required an end to the practice of using favoritism in allocating railroad cars. He urged the railway men to put aside selfish practices and to focus on "helping the men in Europe" by equalizing car supply and thus increasing production.[43]

Supported by Morrow, various government officials, and the editors of *Coal Age*, who now saw Garfield as a "man not easily turned from his course," Garfield got Wilson to approve a memorandum stating that the government "definitely abandons the practice whereby the price of locomotive fuel was reduced in return for a full car supply." Backed by Josephus Daniels, who wanted Wilson to commandeer coal, McAdoo and Williams contended that the operators were making unconscionable profits. Garfield compromised. On May 25, 1918, he ordered that the price of all coal—both commercial and railroad grades—be lowered ten cents a ton. "The President," Garfield stated, "has directed that the railroads

43. Garfield to Wilson, April 10, 1918, Box 92, Garfield Papers; Garfield to Wilson, April 16, 1918; Wilson to Garfield, April 25, 1918; Garfield memorandum to Wilson, May 15, 1918, in Reel 293, Wilson Papers; Garfield telegram to International Railway Fuel Association, May 23, 1918, Records of Harry A. Garfield, Interoffice Memoranda and Communications, RG 67, NA; *New York Times*, March 13, 1919; *Coal Age* 13 (April 20, 1918): 746.

pay the Government price for coal." Although Morrow and others in the Fuel Administration asserted that Williams continued his old practices at times during the summer, the agreement brought a significant improvement in car supply.[44]

Floyd Parsons, the editor of *Coal Age,* glowed over Garfield's leadership. "Settlement of the railroad fuel controversy is regarded here as an unqualified victory for the coal operators," he wrote, "championed as they were by Dr. Harry A. Garfield and his organization." "Many people have believed," Parsons noted, "that the production and national distribution of coal could not be systematized." Yet because of Garfield's zoning and statistical program, "before the snow flies again, Dr. Garfield and his associates will have definite figures showing the production and consumption of coal in every zone, and these data will enable them to shift coal promptly from one region where there is an abundance to other districts where a shortage exists." Equal car supply would insure "that we will have far greater efficiency in coal distribution than ever before." Parsons went on, "We must adapt our lives to a new order of things. . . ."[45] War had definitely created a "new order" for the sprawling, disorganized bituminous industry.

Conservation and Production

Garfield had also been attempting to reorder consumers' habits. He banned outdoor lighting for advertising purposes in the fall of 1917 and instituted "lightless nights" on Thursdays and Sundays in December. On two days a week the famed "Great White Way" in New York was dark. He urged

44. Extensive correspondence on this controversy is in Reel 292, Wilson Papers; FA, *Publication No. 26,* May 25, 1918; "Memorandum for Mr. Garfield," June 25, 1918; Morrow to Garfield, June 26, 1918, Subject File, Fuel Administration, Box 127, Garfield Papers; "Memorandum: Points to be Presented to Operators," with handwritten notes of Harry Garfield, Box 92, Garfield Papers; memorandum, Morrow to Garfield, June 12, 1918, Records of Harry A. Garfield, Interoffice Memoranda and Communications, RG 67, NA; Daniels, *Cabinet Diaries,* April 3, 1918, p. 297; *New York Times,* May 25, 1918; *Economic World* 15 (April 20, 1918): 553; Garfield testimony in *Publication of Production and Profits,* pp. 1292–93; *Coal Age* 13 (April 20, 1918): 723.
45. *Coal Age* 13 (June 1, 1918): 1020; (June 8, 1918): 1043.

Wilson to support daylight-saving legislation. Wilson voluntarily turned off the heat in both the State Dining Room and the East Room. He also refused Garfield's offer to remove the country clubs where he played golf from the establishments denied fuel. In cooperation with the WIB, Garfield denied breweries, greenhouses, cement producers, and window-glass makers full fuel supplies. William Prudden, state fuel administrator for Michigan, got Garfield to close amusement resorts and even industrial plants to save coal for Michigan residents.[46]

Conservation chief P. B. Noyes slowly developed a broad attack on waste. The Fuel Administration's Bureau of Education inundated the nation with pamphlets, posters, questionnaires, movies, speakers, and billboard advertising on saving fuel. Factory committees in over one hundred thousand plants made engineering improvements to conserve fuel. A director of conservation in each state marshaled town and county volunteers to teach consumers to save fuel. Forest rangers showed citizens how to cut wood for fuel without denuding forests. Wisconsinites organized "cut-a-cord" clubs. School children brought home tags with coal-saving suggestions to attach to the family's coal shovel. At Jacksonville, Florida's, "Tag Your Shovel Day" in January, 1918, amid patriotic speeches and "appropriate ceremonies," fifteen thousand persons watched a man tag William Jennings Bryan's axe.[47]

46. Garfield to Joseph Tumulty, February 2, 1918; Garfield to Wilson, November 13, 1917; Garfield to Wilson, July 3, 1918; Garfield memorandum for Wilson, July 13, 1918, all in Reel 293, Wilson Papers. Albert H. Wiggin to Pierpont B. Noyes, December 29, 1917; Noyes memorandum to Mr. Garfield, January 3, 1918; FA, Press Release No. 235; Oliver H. Hewit to Garfield, April 29, 1918; Noyes undated memorandum for Mr. Garfield; all in Records of Harry A. Garfield, Interoffice Memoranda, RG 67, NA. Morrow Memorandum for Mr. Garfield, February 12, 1918; Garfield to Nims, January 4, 1918, in Records of Harry A. Garfield, Interoffice Memoranda, RG 67, NA; McMaster, vol. 2, p. 58; *New York Times*, December 15, 1917; January 1, 1918.

47. Pierpont Noyes to Joseph Tumulty, January 1, 1918; Garfield to Tumulty, December 31, 1917; Program for "U.S. Fuel Administration Mass Meeting, Sunday's Tabernacle, January 28" (1918), all in Reel 293, Wilson Papers. Schmidt and Ault Paper Company to David Moffett Myers, U.S. Fuel Administration, July 15, 1918, in Records of Harry A. Garfield, Fuel Administrator, Interoffice Memoranda, RG 67, NA; *Coal Age* 12 (December 29, 1917): 1100; *Literary Digest*, January 5, 1918, p. 19; Reports of Arthur T. Williams and W. N. Fitzgerald, in FA, *Report of the Distribution Division*, pp. 67, 412; FA, *Final Report*, pp. 246–55.

The conservation program reached out to nearly every aspect of fuel use. Garfield eventually banned general nonessential outdoor illumination and restricted elevators to stops above the third floor. He ordered streetcars to skip stops to conserve the electrical energy used in starting and stopping. Noyes put out a series of pamphlets on engineering techniques in steam and fuel economies. In conjunction with the RA, the Fuel Administration launched a campaign of economy in fuel use in locomotives, repair stations, and roundhouses. Only the armistice halted a gigantic campaign to distribute to every householder in America a window pledge card, signed by Wilson and with a place for the householder to sign, "solemnly agreeing, for himself and household, to save coal in response to the President's appeal."[48]

To avoid hoarding, the Fuel Administration in March, 1918, required licenses for coal dealers and began a rationing program for domestic consumers. Dealers had to secure licenses from the Fuel Administration's Legal Division, which scrutinized applications carefully. On a special form, dealers kept track of each consumer's coal on hand, the amount on order, his requirements for the coming year, and his actual use for the past year. After filing such statements, domestic consumers could purchase only two-thirds of their usual requirements. Through its varied techniques, the Fuel Administration boasted that it conserved 32.2 million tons of coal during its lifetime.[49]

The winter crisis prompted a drive to step up production during the coal year of April 1, 1918, to April 1, 1919. In May the secretary of the navy predicted that naval fuel requirements for the year would treble or quadruple. Shipping Board chief Edward N. Hurley expected to spend an addi-

48. FA, *Final Report*, pp. 246–57. Woodrow Wilson, "To Each American Householder," Garfield to Wilson, September 6, 1918; Noyes to Thomas W. Brahany, chief clerk, White House, August 29, 1918, Reel 293, Wilson Papers. FA, "Saving Coal in Steam Power Plants," "Boiler Water Treatment," "Burning Steam Sizes of Anthracite," "Combustion and Flue Gas Analysis," Technical Papers Nos. 217–222, 1919.

49. FA, *Publication No. 22*, March 18, 1918. Forrest Andrews to John K. Shields, April 12, 1918; Shields to Garfield, April 18, 1918; FA to D. C. Cambell Coal Co., April 20, 1918, all in Records of the Office of the General Solicitor, RG 67, NA. Willoughby, p. 311; FA, *Final Report*, p. 252.

tional billion dollars to increase shipping. Morrow estimated that current rates of coal production would leave the country 71 million tons short for the coal year. When production fell off during the week of April 1, the miners' traditional celebration of the anniversary of the eight-hour day, Garfield predicted the winter would bring another coal famine.[50]

So, on June 7, 1918, Garfield established a new Bureau of Production headed by James B. Neale. Garfield's advisor on anthracite, Neale was an anthracite operator and a Yale graduate who had been recommended by Vance McCormick, head of the War Trade Board. Neale organized a moralistic propaganda drive to stimulate output. Crippled soldiers like Private John F. Hannon, who had lost an arm in France, addressed rallies in the coal fields, often in the language of the miners' native lands. Movies, posters, banners, and letters from leading civic and religious figures encouraged greater output. Retired miners volunteered to return to the pits. To cut lost time, picnics and other celebrations in the mining camps were cancelled. Even funerals for miners no longer involved an entire community; by the end of the war only a committee of miners would attend a mine-camp funeral.[51]

Under Neale's direction, the Fuel Administration's cooperative thrust embraced miners and operators in individual mines. Neale appointed special production managers for each district who set quotas and oversaw joint miner-operator production committees for each mine. Composed of three miners elected by their fellow workers or union local and three representatives of management, these production committees boosted output. With the NCA's full support, miners on the production committees had the right to criticize management decisions which retarded production. Wearing spe-

50. *New York Times*, April 16, May 29, June 11, 1918; Daniels to Garfield, May 29, 1918, Records of Harry A. Garfield, General Correspondence, RG 67, NA. Cyrus Garnsey, assistant U.S. fuel administrator, to P. B. Noyes, August 6, 1918, in Records of Harry A. Garfield, Interoffice Memoranda, RG 67, NA.

51. Garfield to Daniels, March 3, 1921, Box 86, Garfield Papers; *Shortage of Coal*, p. 696; FA, *Final Report*, pp. 243–45; *Coal Age* 14 (August 29, 1918): 405; Baker to Garfield, July 30, 1918, Records of Harry A. Garfield, General Correspondence, RG 67, NA; *United Mine Workers Journal*, July 18, 1918, pp. 5, 16; August 15, 1918, p. 12.

cial badges, the mine committeemen also coordinated the propaganda campaign, kept after laggards, and even—where patriotism was in full flower—pressured liquor dealers to "restrict abusive sales of liquor" to miners. After debate with the secretary of war, the Fuel Administration won deferred military classification and special furloughs for essential mine workers. Neale also kept after the RA to insure that the car-supply agreement was carried out.[52]

In August, 1918, Wilson issued a special proclamation to the industry and its workers. Noting that scarcity of coal was "creating a grave danger," he lauded the operators for trying to improve working conditions and management techniques. Yet he lectured the miners: they should report to work unless "prevented by unavoidable causes." They should stay on the job all day and try to get out more coal. He said that those receiving deferments should accept them as a "patriotic duty," and communities should not denigrate those deferred. "The only worker," said Wilson, "who deserves the condemnation of his community is the one who fails to give his best in this crisis."[53]

Under the lash of committeemen, quotas, propaganda, improved car supply, and the reduction in cross-hauling brought by zoning, production boomed. The Connellsville District, which specialized in coal for coke, worked the Fourth of July and produced enough coal to supply the energy to produce the steel to manufacture twenty average-sized cargo ships. During the week including Labor Day, Connellsville production hit its highest weekly total for the year. Despite the massive influenza epidemic which curtailed working time in all industries during late October, total production for the month of October was four and one-half

52. FA, *Final Report*, pp. 243–45; William Hard, "Coal," *New Republic* 16 (September 21, 1918): 228; Garfield to Wilson, January 9, 1918, with plan for deferred classification of miners, Box 8, Baker Papers. Adam B. Littlepage, congressman, West Virginia, to Tumulty, September 10, 1918; Garfield to Tumulty, January 3, 1918; Garfield to Wilson, January 9, 1918; Neale memorandum, December 28, 1917, all in Reel 293, Wilson Papers; McMaster, vol. 2, p. 59; *United Mine Workers Journal*, July 25, 1918, p. 8; August 8, 1918, p. 9; September 15, 1918, p. 8; November 1, 1918, p. 8.

53. Neale to Garfield, July 31, 1918, Reel 293, Wilson Papers; Lane to Wilson, August 6, 1918; Wilson to Lane, August 8, 1918, Reel 289, Wilson Papers; *Coal Age* 14 (August 15, 1918): 319.

million tons ahead of the same month a year before, as was every month to the armistice under the "new order of things." Despite the dire forecasts of the summer, by the end of the war stocks were at record levels. War demand and effective business-government cooperation had reversed the economic position of a traditionally "sick" industry.[54]

Garfield succeeded by relying on the talents of the industrialists. Following the January crisis at the ports, he brought operators and skilled personnel into the administration. "The time was," William Hard wrote, "when the Fuel Administration was not conspicuous for personal talent or experience in fuel. It seemed to be built on the distinctively American principle that regulatory bodies should be as alien as possible from the technique of the field to be regulated." But by mid-1918, Hard continued, Garfield was "surrounded by great quantities of thoroughly experienced technical advice and assistance." *Coal Age* praised the Fuel Administration officials as "capable, patriotic men who are giving unselfishly the best that is in them." Hard complimented both Garfield and the "patriotism and good business judgment of the members of the National Coal Association, who have volunteered, at great immediate cost of private money and of private time, to help steer the boat instead of being content with rocking it."[55]

War demand had given the operators a short-run bargaining position. No longer at the mercy of large oligopolistic buyers, the industry could afford zones. Because coal was the nation's most important energy source and vital to winning the war, Garfield could demand a better car supply. Once the government took over the railroads and improved transportation, coal production soared. War unified the objectives of the industry and the government, softened the harsh edges of competition, and paved the way for coopera-

54. "Conference Held in Private Office of Mr. Garfield," July 30, 1919, Records of Harry A. Garfield, General Correspondence, RG 67, NA; FA, *Final Report*, pp. 21–22, 243–45.

55. Hard, "Coal," pp. 226–28; Hard, "Coal—Why Not More?" *New Republic* 16 (October 5, 1918): 276–77; *Coal Age* 14 (July 13, 1918): 47; Interstate Commerce Commissioner Robert Wolley to Tumulty, September 23, 1918, Reel 293, Wilson Papers.

tive effort between business and government. *Coal Age* wrote that "great progress has been made toward the orderly and effective control of an immense industry of vital importance."[56] But this attitude did not maintain itself where industry objectives and government demands diverged.

Wages, Prices, and Collective Bargaining

In handling wages, prices, and collective bargaining, Garfield tried to balance competing interests. To maintain production, he backed union demands for grievance machinery and recognition of collective bargaining where locals existed. In return, the UMW leadership, which now included John L. Lewis as international statistician, agreed to moderate its unionization campaign in the South. During the war years the UMW in fact boosted its membership to the highest levels up to that time and with a claimed four hundred thousand members was the largest union in the country. Garfield's committee of engineers adjusted prices upward in cases where union wage gains threatened to raise production costs above a break-even point.

In the summer of 1917 the UMW sought to renegotiate the 1917 wage contract in which the UMW and Central Field operators in New York had agreed to a wage increase of ten cents a ton and sixty cents a day for day laborers, effective April 16, 1917. But inflation continued; operators made profits; some mine workers left for other higher-paying industries; and in late summer, wildcat strikes increased the pressure on the operators and the Fuel Administration for renegotiation of the Central Field contract. The newly appointed fuel administrator urged that operator-miner discussions be moved from Indianapolis to Washington, D.C.[57]

56. *Coal Age* 14 (July 13, 1918): 47; Van Hise, p. 169. Louis Galambos, *Competition and Cooperation: The Emergence of a National Trade Association* (Baltimore: Johns Hopkins University Press, 1966), pp. 64–68; Rex Tugwell, "America's War-Time Socialism," *Nation* 124 (April 6, 1927): 364–66.

57. Melvin Dubobsky and Warren Van Tine, *John L. Lewis* (New York: Quadrangle/The New York Times Book Co., 1977), pp. 34–40; Charles B. Fowler, *Collective Bargaining in the Bituminous Coal Industry* (New York: Prentice Hall, Inc., 1927), p. 79; *New York Times*, May 1, 1917; *Coal Age* 12 (September 29, 1917): 539.

Garfield told the reassembled wage conference on September 25, 1917, that "under no circumstances" could the administration allow coal production to stop. Under government pressure, and because operators knew they could pass wage advances on in higher goverment prices, they granted another increase. Coal diggers got $0.10 a ton more, and day laborers got raises of between $0.75 and $1.40. In exchange for total increases for the two renegotiations of roughly 50 percent over the 1914 levels, the UMW spokesmen agreed to district ratification of a penalty clause for striking during the war. Effective "only on the condition that the selling price of coal shall be advanced" enough to pay the new wage, this Washington Agreement was to run (in a phrase which became enormously significant later) "during the continuation of the war, and not to exceed two years from April 1, 1918."[58]

Technical advisor Samuel A. Taylor and Garfield decided that a forty-five-cent increase in Wilson's prices would cover increased costs brought by inflation and the wage agreement. Shortly after the wage conference broke up, Garfield took his figures and an FTC study of the situation to a meeting with the president. Reported to be still disappointed by the CCP's attempt to peg prices at three dollars, and reluctant to alter his own prices, Wilson agreed to the general increase of forty-five cents a ton effective October 29, 1917.[59]

The Washington Agreement stabilized wages for the next year, but Garfield had to intervene in several other labor disputes during the war. Following a long strike in Alabama, he convinced operators there to accept a fourteen-part decision to recognize the right of their employees to "join any union, labor organization, or society they may choose" and

58. Fowler, pp. 80–81; *Coal Age* 12 (September 29, 1917): 539–41; (October 13, 1917): 635; Willoughby, 297–99; Report of W. J. Galligan, in FA, *Report of the Administrative Division*, pp. 34–35; Garfield testimony, U.S., Congress, Senate, Committee on Interstate Commerce, *Increased Price of Coal: Hearings Pursuant to S. Res. 126*, 66th Cong. 1st sess., 1920, p. 518; *New York Times*, October 17, 18, 21, November 4, 7, 20, 1917.

59. Daniels, p. 211; *New York Times*, September 29, October 3, 28, 1917; Glen Lawhon Parker, *The Coal Industry: A Study in Social Control* (Washington, D.C.: American Council on Public Affairs, 1940), p. 87; *Coal Age* 12 (September 22, 1917): 496; (October 13, 1917): 637; (October 27, 1917): 764.

to refrain from discriminating against union workers. The operators further agreed to permit peaceful assembly, to receive grievances from workers' committees, to reemploy men fired for union organizing, and to permit checkweighmen. Garfield forced union leaders and management to work out a satisfactory wage agreement for the Alabama field which lasted until the armistice.[60]

In the Maryland and upper Potomac districts, the fuel administrator intervened in another dispute which threatened to disrupt production. He summoned the operators' and miners' spokesmen to Washington and got the operators to agree to carry out the principles established in Alabama. The contract spelled out a full labor-management agreement for grievance procedures, including election of mine committeemen, right of appeal to an umpire appointed by a federal judge, whose decision could be reviewed by the Fuel Administration, and election of checkweighmen. War profits apparently greased the wheels of collective bargaining just as they had encouraged operators and the government to cooperate.[61]

As his advisor on labor Garfield had appointed former UMW President John P. White. Rembrandt Peale, a central Pennsylvania operator who had helped initiate the Tidewater Coal Exchange during the summer of 1917, counseled Garfield on bituminous coal. In the spring of 1918 Garfield brought the two together as heads of a Fuel Administration Bureau of Labor. Working with the new bureau and representatives of the UMW, Garfield issued in May a balanced memorandum codifying Fuel Administration policy on labor disputes. It demanded that (1) there be no strike until a controversy had been reviewed by the Fuel Administration, (2) "recognition of the Unions shall not be exacted during the continuance of the war except where now recognized," (3) grievance machinery was to be employed where it existed, and (4) where Garfield had to intervene, the princi-

60. Samuel Gompers to Wilson, December 11, 1917; Wilson to Gompers, December 12, 1917, Reel 292, Wilson Papers; FA, *Final Report*, pp. 211–12; *Coal Age* 13 (January 19, 1918): 152–53.

61. FA, *Final Report*, pp. 212–14.

ples of the Maryland and upper Potomac settlement would apply. Further, employers relinquished the right to discriminate against union members and recognized the right of their employees to organize. In exchange for these benefits the union leaders had to accept the automatic penalty clause for striking.[62]

Garfield thus steered a middle course. By both supporting grievance machinery and moderating union demands, the Fuel Administration reduced time lost to strikes and stoppages during the war. Wilson's prices retarded production and angered the industry, but the forty-five-cents-a-ton increase helped to mollify operator resentment.[63]

Following the president's announcement of the two-dollar base price in August, 1917, operators descended on Washington to seek exceptions—some even before Garfield had been appointed. Although Wilson had relied upon FTC data for his prices, and despite the Lever Act's statement that "the commission" shall determine prices, Wilson gave Garfield the job of revising the temporary August prices. Garfield forwarded most of the initial complaints to H. D. Nims's legal department but soon appointed Edmond Trowbridge to head a Bureau of Prices. Working out of an unheated butler's pantry on Sixteenth Street, where the Fuel Administration had temporary offices, Trowbridge brought in FTC accountants to recommend revisions for Garfield to approve. Using the FTC cost data for August, 1917, and in cooperation with commission accountants, Trowbridge, Nims, and Garfield initially revised individual prices on particular kinds of coals.[64]

62. Garfield to Tumulty, September 13, 1917; telegram from Tumulty to Wilson, September 13, 1917; Wilson to Daniels, September 14, 1917, Reel 293, Wilson Papers; *Coal Age* 12 (September 22, 1917): 491; (October 6, 1917): 579; *United Mine Workers Journal*, December 1, 1917, p. 7; Frank J. Hayes to Garfield, July 20, 1918; Garfield to Hayes, July 15, 1918; "Memorandum of Statement of Principles made by Mr. H. A. Garfield, at Conference Held May 16–18, 1918," Records of Harry Garfield, General Correspondence, RG 67, NA; Garfield Memorandum, August 29, 1919, Box 93, Garfield Papers.

63. Frank J. Hayes, John L. Lewis, and William Green to Garfield, July 31, 1918, Records of Harry A. Garfield, General Correspondence, RG 67, NA; reports of Thomas W. Russell and H. J. M. Jones in FA, *Report of the Administrative Division*, pp. 46, 366; John M. Clark, *Social Control of Business* (New York: McGraw-Hill Book Co., Inc., 1939), p. 242; Cuff, *War Industries Board*, pp. 220–41.

64. Lane Moore, National Archives staff member, shared with me the research he has done on the FTC coal investigation, particularly the correspondence and

But the Bureau of Prices needed to develop a systematic technique for determining specific price revisions. The Lever Act mandated that the government prices "shall allow the cost of production, including the expense of operation, maintenance, depreciation, and depletion, and shall add hereto a just and reasonable profit." Garfield appointed Cyrus Garnsey as head of an engineers' committee to assist Trowbridge in developing a system for determining revisions for particular kinds of coals in the eighty-one Fuel Administration districts of the country. Garnsey, a portly man who had been involved in mining and railroading and had the confidence of both operators and miners, left a quiet retirement in Seneca Falls, New York, to become a dollar-a-year man for the Fuel Administration. To assist Garnsey, Garfield brought in two mining engineers, R. V. Norris and James Alport.[65]

The committee rejected straight cost plus for each operation because it led to "multiplicity of prices," potential changing prices as costs altered, and possible reduction in output as operators mined their poorer coal at high costs and saved their better coal for the war's end. Average cost plus for each district, they believed, would fail to draw out the production of the higher-cost mines. Pooling they thought to be too cumbersome, too expensive, and too much of an innovation to be attempted during the crisis of war.[66]

The engineers' committee finally worked out what they termed a "bulk line" method of determining acceptable revisions. Using the operators' yearly cost statements for 1916 and the monthly ones for August and September, 1917, sub-

memoranda between the commission and the Fuel Administration. Nims, "Memorandum Regarding Production," November 20, 1917; Nims, "Memoranda for Dr. Garfield," October 2, 22, 24, 1917, Records of Harry A. Garfield, Interoffice Memoranda, RG 67, NA. The actual statutory powers over price fixing were not transferred from the FTC to the Fuel Administration until July 3, 1918. *Coal Age* 14 (October 10, 1918): 685; Garfield testimony in *Shortage of Coal*, p. 7; FA, *Final Report*, pp. 240–43.

65. Garfield to Daniels, March 3, 1921, Box 86, Garfield Papers; J. H. Bankhead to Garfield, May 16, 1918, Reel 293, Wilson Papers; *Coal Age* 14 (October 10, 1918): 685; David L. Wing, "Cost, Prices, and Profits of the Bituminous Coal Industry," *American Economic Review* 11 (March, 1921): 74–78; FA, *Final Report*, pp. 240–42.

66. FA, *Report of the Engineers Committee* (Washington, D.C.: Government Printing Office, 1919), pp. 1–5; Norris testimony, Senate Committee on Interstate Commerce, *Increased Price of Coal*, p. 566; Garrett, p. 169.

mitted to the FTC, the engineers developed cost charts for each coal district. These charts graphed a district's costs, showing which percentage of output was produced at which cost. With these charts of the eighty-one districts they determined what price was needed to maintain production of the "bulk" of the tonnage for the district. Although the highest-cost mines in a particular district might not profit under the revised price, usually over 90 percent did, and the lowest-cost operations would profit handsomely.[67]

After Garfield tested the engineers' system on the Piney Fork District in Ohio, where he knew costs at first hand, he cleared the plan with Wilson. The engineers' system produced high profits for low-cost mines, but FTC field agents checked the data behind any revision. The engineers denied many applications for upward revision and lowered the prices for some districts, even in the face of complaints from one senator. Price adjustments typically involved creating subdistricts within districts to account for higher regional production costs. The engineers often forced operators to reduce charges for royalties paid but also told operators that they should include depreciation in their costs. Operators who violated the Fuel Administration prices were investigated and fined. Trowbridge boasted of having taught the industry how to keep accurate cost accounts.[68]

67. U.S., Federal Trade Commission (henceforth cited as FTC), "Summary Interim Report" (typewritten), Reel 308, Wilson Papers; Garfield to Wilson, February 11, 1918; Wilson to Garfield, February 13, 1918, Reel 293, Wilson Papers; *Shortage of Coal*, pp. 38–39; *Coal Age* 13 (June 1, 1918): 1020; FTC, *Cost Reports: Coal, Nos. 1–9*, 1919; U.S., Congress, House, Select Committee on Expenditures in the War Department, *Expenditures in the War Department*, 66th Cong., 1st–3rd sessions, 1919–21, serial 3, vol. 2, p. 2696; see FA, *Report of the Engineers Committee*, published also in *Coal Age* 14 (October 10, 1918): 685–89; (October 17, 1918): 736–39.

68. There are complete correspondence records on the thousands of applications for revision in Records of the Bureau of Prices, RG 67, NA. FA, "Morning Bulletin" (mimeographed), July 24, 1918, Subject File, Fuel Administration, Garfield Papers; Memorandum, July 1, 1918; Garfield to Senator Charles S. Thomas, March 20, 1918, Reel 293, Wilson Papers; Garfield testimony in *Publication of Production and Profits*, pp. 1288–89; Garnsey, Norris, and Allport to Garfield, March 20, 1918, Records of Harry A. Garfield, Interoffice Memoranda, RG 67, NA. Investigators toured forty-eight mines in Ohio to determine the proper distinction between thick-vein coal and thin-vein coal. J. H. Pritchard, Commissioner of Mining, Southern Ohio Coal Exchange, to Francis Walker, Chief Economist, Violations Division, Bituminous Coal Section, FTC, July 15, 1918, and Investigation 9165 in Records of the Bureau of Investigation, RG 67, NA; *Coal Age* 13 (February 2, 1918): 251.

Fuel Administration maximum-price controls for the first seven months of operation—to the beginning of the 1918 contract year in April—kept the weighted average price on all bituminous to $2.17 per ton, $0.19 below the weighted average price of $2.36 per ton for the earlier five months of that coal year. This reduction represented a theoretical savings of some sixty-one million dollars for that period alone. It thus can be argued that in the initial stages, the Fuel Administration effectively lowered prices and saved consumers money.[69]

During the last eight months of the war, average prices increased. First, because the April, 1918, contracts included the fall wage and price increases, the weighted average price jumped $0.51 a ton. Then the across-the-board reduction of $0.10 the Fuel Administration agreed to in the May, 1918, negotiation with the Railroad Administration lowered the weighted average to $2.66 a ton for the duration of the Fuel Administration controls in 1918. The weighted average prices under the new contracts were thus $0.32 above those of the precontrol months of 1917, a theoretical additional cost of $42 million. Together, the two periods of regulation, compared with the months of 1917 when there were no controls, brought theoretical savings of $19 million. Because the Fuel Administration prices averaged $0.388 less than those proposed by the CCP, which the administration nullified, some have concluded that the Fuel Administration saved consumers as much as $312 million.[70]

69. Garrett, pp. 24, 153, 170. Charles Van Hise estimated that without the price controls, coal would have been at least two dollars higher during the war (p. 168); FA, *Final Report*, p. 20. Spot prices for European coal shot up to sixty and eighty dollars a ton during the war (Palmer, vol. 2, p. 30). Ellery B. Gordon and William Y. Webb, *Economic Standards of Government Price Control*, TNEC Monograph 32 (Washington, D.C.: Goverment Printing Office, 1940), p. 247; George P. Adams, *Wartime Price Control* (Washington, D.C.: American Council on Public Affairs, 1942), pp. 18, 143–47.

70. Garrett, p. 171; Parker, p. 86; Wing, "Cost, Prices, and Profits," p. 82, Table 7; Bentley Warren, "Memorandum as to the Bituminous Coal Situation," February, 1919, Box 93, Garfield Papers; FTC quoted by *Coal Age* 14 (July 6, 1918): 21; FTC, *Cost Reports*, pp. 4–5; Senate Committee on Interstate Commerce, *Increased Price of Coal*, pp. 611–13; Garfield testified to his own company's profits in *Publication of Production and Profits*, pp. 1291–92; see also pp. 1559–67; Gordon and Webb, p. 247. Returns on investment for U.S. Steel

Although aggregate demand might not have kept prices at the CCP levels throughout the war, as the operators had argued, America's wartime need for energy was immense. The United States government and the Allies spent some $31.5 billion on war goods and services in the United States during the war. The National Bureau of Economic Research estimated that American war expenditures alone constituted one-fifth of the national income in 1917 and one-fourth of the national income in 1918. Coal users formed the backbone of the war economy. Railroad transportation, dependent on coal, went up 24 percent from the 1914 level in 1916, up 36 percent in 1917, and up 42 percent in 1918. Steel output doubled in 1917 over the prewar norm and stabilized at that level for 1918. Spending on plant and equipment jumped from $600 million in 1915 to $2.5 billion in 1918. Energy consumption rose 32 percent between 1914 and 1918. Coal's share of the energy market was still approximately 70 percent (see Fig. 2, p. 21). In 1918, per capita consumption of coal reached its highest level in American history.[71]

The data in Table 1 show how new mines sprang up to meet war demand. The industry's mines also worked more days during the war. But because railroad cars were in short supply, production even during war did not reach full capacity (see Fig. 1, p. 20). Even during the war, the bituminous industry did not go to two shifts. It increased mechanized cutting by 4 percent between 1914 and 1918. Working 308 days a year, the industry could have produced 700 million

---

climbed from an average of 4.6 percent for 1912–15 to 15.6 percent in 1916 and 24.9 percent in 1917; one of the nation's two sulfur producers earned 236 percent return on investment in 1917; in the 1917 lumber industry the FTC found that "the range of profits was from a small loss to over 121 percent on net investment." Between 1915 and 1917, profits for the big four meat packers increased 400 percent. FTC, "Report on Profiteering, 1918," in U.S., Congress, House Committee on Military Affairs, *Hearings on Taking the Profits out of War*, 74th Cong., 1st sess., 1934, pp. 604–16.

71. George Soule, *Prosperity Decade: From War to Depression, 1917–1918* (New York: Holt Rinehart and Winston, 1968), chapter 2; U.S. Bureau of the Census, *Historical Statistics of the United States: Colonial Times to 1957* (Washington, D.C.: Government Printing Office, 1960), p. 355; Harold Barger and S. H. Schurr, *The Mining Industries, 1889–1939: A Study of Output, Employment, and Productivity* (New York: National Bureau of Economic Research, 1944), p. 74.

Table 1. Operating Mines, Working Days, and Production, 1916–1919.

	Operating Mines	Percentage of Increase or Decrease	Days Worked	Percentage of Increase or Decrease	Production (Millions of Tons)	Percentage of Increase or Decrease
1916	5,726		230		503	
1917	6,939	+21	243	+5.6	552	+9.7
1918	8,319	+20	249	+2.4	579	+4.9
1919	8,994	+8	195	−21.6	466	−19.6

SOURCES: Ellery B. Gordon and William Y. Webb, *Economic Standards of Government Price Control*, TNEC Monograph 32 (Washington, D.C.: Government Printing Office, 1940), chap. 1, and U.S., Bureau of the Census, *Historical Statistics of the United States: Colonial Times to 1957* (Washington, D.C.: Government Printing Office, 1960), p. 356.

tons, even on a single shift. This capacity could have been reached, however, only with even higher prices and a full supply of railroad cars.[72]

Prior to regulation the industry had experienced its 1916–1917 boom. Then the administration attacked the CCP's attempt to impose maximum prices. This opposition angered coal men and hindered production during September. But the boom resumed, even under Wilson's prices. The Fuel Administration prices were lower than equilibrium and saved consumers money, yet after incorporating the costs of the Washington Agreement, they were still high enough to encourage record production. But they were not high enough to induce the reallocation of rail or labor resources necessary to reach the industry's hypothetical full capacity.

Data on income and return on equity show that the industry earned more in 1917, when it was unregulated for nine months, than it did in 1918 under controls. Unfortunately, coal industry fiscal years and Internal Revenue Service fiscal years do not correspond with coal contract years and make comparison of yearly prices and earnings difficult. Moreover, there are no reliable aggregate figures for return on capital or profit for the industry except the Internal Revenue Service data. These statistics begin in 1917 for the coal industry, are aggregate figures, and, given the number of small, family mines which do not report, are not very reli-

72. U.S., Federal Works Agency, WPA National Research Project, *Mechanization, Employment, and Output per Man in Bituminous Coal Mining* (Washington, D.C.: Government Printing Office, 1939), pp. 96–97.

able except in showing large-scale changes in the industry's fortunes. The Internal Revenue Service shows net income for the industry in 1917 to be $203,919,000, in 1918 at $148,847,000, and in 1919 at $62,260,000 (see Fig. 1, p. 20). The 1922 U.S. Coal Commission reported the before-tax annual rate of return on equity for a sample of eighty-eight companies. The average rate for the 1913–14 base was 4.15 percent. It was 3.9 percent in 1915. In 1916, it grew to 9.1 percent, reached a peak in 1917 of 24.8 percent, and then fell back to 7.5 percent in 1918 and 7.7 percent in 1919.[73] The industry's peak profit year was 1917, a year before price controls affected the bulk of coal sales.

Thus, as was true for prices, the wartime regulation of the industry allowed it better than 1913–14 rates of return on equity but apparently kept rates lower than they would have been without regulation, given the impact of wartime demand for coal. The Fuel Administration calculated that the industry's wartime 45.6-cent "margins"—the difference between average costs and average selling prices—were double those of 1916, but those of 1918 were lower than those of 1917. Whether the Fuel Administration controls prevented what the *United Mine Workers Journal* called "the carnival of criminal extortion that would have prevailed in the absence of regulation," no one will never know. The New Deal's Temporary National Economic Committee study of the bituminous industry concluded that there was little indication of profiteering under Fuel Administration controls.[74]

Regulation under the Fuel Administration aimed, of course, not only at reducing excess profits but also at increasing production. The industry paid larger-than-normal income

73. Inquiries to *Coal Age*, the National Coal Association, the Internal Revenue Service, coal economists, and security analysts failed to turn up any source for dependable return-on-investment figures for the coal industry as a whole. Individual corporation annual reports are very difficult to summarize, since coal company earnings derive from a variety of sources other than coal mining, and there was no standard system of reporting earnings. Gordon and Webb, p. 243; U.S., Coal Commission, *Report*, 68th Cong., 2d sess., 1925, pp. 2517–24, 2677–83; FTC, *Preliminary Report on Investment and Profit in Soft-Coal Mining*, 1922, pp. 3–12.

74. *United Mine Workers Journal*, March 1, 1919, p. 6. Gordon and Webb, p. 247.

taxes on excess profits under the war revenue acts. The Lever Act, a progressive triumph under which Wilson lowered spot coal prices, mandated that government prices allow a "just and reasonable profit." Garfield and his engineers regulated prices to "secure the last necessary ton of coal for each of the several districts." Setting the bulk line high enough to produce coal from some of the highest-cost mines insured profits to all but marginal operators. Considering the magnitude of the project of stabilizing the coal industry, the success of the FTC and the Fuel Administration in controlling prices while encouraging production and streamlining distribution was a significant achievement.[75]

Wilson announced on November 11, 1918, that the armistice had been signed and that "everything for which America fought has been accomplished." Thousands celebrated. But as whistles blew on mine tipples and industrial plants, buyers of war materials jammed switchboards with order cancellations. Within a month the War Department retracted half of its outstanding contracts. Soon after the crowds had trampled chalk drawings of the Kaiser and college girls had finished the snake-dance celebrations of the war's end, dramatically reduced demand for coal brought falling prices, particularly for poorer grades. War maximums became meaningless. War administrators yearned to return to private life. The demobilization of the Fuel Administration began.[76]

Although he would be drawn back into the industry's postwar problems, including the wage controversy that resulted in the famous strike of 1919, in November, 1918, Garfield longed to return to Williams College. He submitted his resignation to Wilson in early December, 1918. The presi-

75. Lever Act, 40 Stat. 285 (1917); Garfield testimony in *Shortage of Coal*, p. 12; FTC, *Cost Reports*, vol. 1, p. viii; *Coal Age* 14 (December 5, 1918): 1054; report of S. P. Kennedy, in FA, *Report of the Administrative Division*, p. 14; FA, *Final Report*, p. 24.

76. Frederick Lewis Allen, *Only Yesterday* (New York: Bantam Books, 1957; originally published by Harper & Brothers, 1931), pp. 11–12; Soule, p. 81; Morrow testimony, in Senate Committee on Interstate Commerce, *Increased Price of Coal*, p. 126; Robert Cuff in *War Industries Board*, chapter 9; Robert F. Himmelberg, "The War Industries Board and the Antitrust Question in November 1918," *Journal of American History* 52 (June, 1965): 59–74; Robert K. Murray, *The Harding Era: Warren G. Harding and His Administration* (Minneapolis: University of Minnesota Press, 1969), pp. 73–74.

dent accepted Garfield's decision to return to Williams and complimented him on "a splendid piece of work." Garfield organized an "entertainment" at Washington's Central High School for his rapidly diminishing staff. On February 1, 1919, convinced that zone and price regulations could be removed without fear of high prices or shortages, Garfield suspended, but did not rescind, the orders establishing both prices and zones. Although it would take until May for the various divisions to pack their records and prepare their reports, the ongoing work of the Fuel Administration ended by the beginning of the new year.[77]

Conclusions

Viewed as a "trust," the coal men were excluded from the Fuel Administration during its early days. Progressive Wilsonians replaced a system of "industrial self-government" under the CCP with a program of tight federal control. But the weather and the abrupt cancellation of the cooperative phase led to shortages and cries that the Fuel Administration needed business talent, which it finally employed in the distribution program. Because coal executives brought their expertise into government and because their cooperation with Garfield produced the rationalization of the industry that the Central Field operators had been unable to achieve in the prewar years, some could argue that the Fuel Administration experience proves once again the general notion that industry tends to influence governmental regulatory agencies created by liberal reformers.

Garfield might thus be labeled a "corporate liberal" who moderated unruly radical elements among the miners and faced down some of the more reactionary anti-union southern operators to bring about a stabilized, more national industry. He helped both the Central Field operators and the UMW leadership to accomplish their private goals through

77. Memorandum, December 3, 1918; Garfield to Tumulty, December 9, 1918, Reel 293, Wilson Papers. Wilson to Garfield, December 3, 1918, Box 93; Garfield diary, January 15, 1919; typescript copy of the suspension announcement, January 15, 1918, Fuel Administration, Miscellaneous, Garfield Papers. Willoughby, pp. 314–15; *Coal Age* 14 (November 14, 1918): 305; (December 12, 1918): 1083.

government. Garfield and the leading Central Field operators spoke the same language and together operated as a
kind of "functional elite" in Kolko's sense of that term. Too,
the organization-minded coal executives that were brought
into the Fuel Administration influenced its operation, provided Garfield with information to lead the fight for equitable car supply, and made the day-to-day operation of the
Fuel Administration more efficient.[78]

In this regard, the coal industry's experience parallels that
of the cotton textile and lumber industries during the war:
both achieved a level of stabilization they could not have
reached during the prewar years. But neither they nor the
soft coal industry exercised the influence on government
during the war that more oligopolistic industries did. Previously stabilized industries like steel and copper, led by U.S.
Steel's Judge Elbert Gary and Anaconda's John Ryan, could
drive a much harder bargain with government over prices,
for example, than could the coal industry. The coal industry
was too fragmented and too highly competitive to speak as a
unit or to truly exploit its position within government. The
cooperation and stabilization the industry achieved during
the war came as a by-product of mobilization for war objectives and at a cost to particular firms such as Morrow's own
Pittsburgh Coal Company, which was denied access to its
traditional midwestern markets under the zone plan. A stub-
born and strong Garfield harnessed a fragmented industry to
the war goals. He was not manipulated by any industry figure, nor did the industry capture the Fuel Administration in
some sinister sense.[79]

Moreover, Wilson's attack on the Peabody committee and
his imposition of prices insured that the coal industry entered government's halls as second-class citizens. They pro-

78. See James Weinstein, *The Corporate Ideal in the Liberal State: 1900–1918*
(Boston: Beacon Press, 1968), chapters 1 and 8, and Gabriel Kolko, *The Triumph of
Conservatism* (Glencoe: Free Press, 1963), conclusion.

79. Cuff, *War Industries Board*, pp. 75–81, 128–29, 164; Galambos, pp. 41, 64–
67; Gerald D. Nash, *United States Oil Policy, 1890–1964: Business and Government in Twentieth-Century America* (Pittsburgh: University of Pittsburgh Press,
1968), chapter 2; Melvin I. Urofsky, *Big Steel and the Wilson Administration: A
Study in Business-Government Relations* (Columbus: Ohio State University Press,
1969).

vided expertise, but they functioned in restricted phases of the Fuel Administration and within limits previously set. They cooperated with government and government with them because the economic conditions of war made cooperation seem imperative and not harmful to profits. But no coal man could speak for a corporate entity that controlled more than 5 percent of the industry. Thus, even given an entrée into government after January, 1918, coal executives did not control government so much as cooperate with the patrician Garfield, and largely on his terms.

But for a brief period, under the direction of Harry A. Garfield the industry experienced a coordination and spirit it had never seen before. Garfield was not alone in hoping that such a cooperative endeavor could be made permanent. After the armistice he called leading industrialists and union representatives together to discuss a dramatic restructuring of America's political economy in order to maintain the cooperative spirit gained by wartime governmental intrusion into the coal industry. "With our entrance into the war we had an object in which everybody believed," he said, "and each of the three parties here represented, namely, the public, capital invested in the industry, and the workers in the industry, shared in the desire that the object be accomplished, hence we were able to cooperate. Now peace is close upon us—are we to return more or less definitely to the old way?"[80]

80. See FA, *Final Report*, pp. 136–67.

4

From Riches to Rags

IN THE 1920's the soft coal industry went from riches to rags and reverted to its cutthroat traditions. Harry Garfield's attempt to maintain business-government cooperation collapsed when the United Mine Workers of America, led by John L. Lewis, struck to catch up with war inflation. Then shortages led to high prices and congressional investigations of coal's profits and production. Becoming increasingly non-union, the National Coal Association and the industry in general turned away from business-government cooperation during the decade.

In 1924 the UMW won a good contract from the Central Field operators in the Jacksonville Agreement. But high wages in the Central Field only further encouraged the growth of non-union mines in the South and exacerbated the industry's age-old problem of "too many mines and too many miners." Demand for coal slacked off, other fuels eroded markets, and excess supply depressed coal prices. Weakened by wage cuts and layoffs, the UMW shrank, and the industry lost its profits. Non-union operators gained control of the NCA. The Great Depression administered the *coup de grâce*. With some support from the organization-minded operators from the Central Field, the UMW—now the only spokesman for the entire industry—devised a special coal bill to protect collective bargaining, wages, and prices. As president, Hoover refused to sanction price-fixing arrangements, and the operators in non-union sections of the industry attacked Lewis's stabilization schemes. The Central Field operator–UMW alliance would have to wait until the administration of Franklin D. Roosevelt for legislative relief for the depressed coal industry.

Thus, despite the successes at stabilization brought by the Fuel Administration, once the war crisis was over, coal operators returned to their traditionally hostile view of government. Although Senator William Kenyon and Secretary of Commerce Herbert Hoover attempted to find a way to stabilize the industry, many in government slid back into seeing the coal industry as a "trust" or "interest" which must be investigated rather than assisted. The growth of the southern fields further eroded what little operator unity there had been during the war. The NCA no longer spoke for the entire industry. Fragmented north and south, and becoming fiercely competitive, the industry once again lacked positive leadership. Ironically, the one national voice to emerge was that of John L. Lewis, who took over the United Mine Workers of America and eventually laid the groundwork for the stabilization measures of the New Deal.

Wilson's Wartime Cooperation Collapses

The war economy eventually began to strain relationships between Garfield and the UMW leadership. Inflation and wage gains in anthracite mining and shipbuilding gave bituminous miners an incentive to renegotiate the Washington Agreement. Even bonus payments to keep miners in the pits could not prevent the diggers from feeling that they were falling behind other laborers. In August, 1918, the UMW demanded a wage increase. Garfield agreed to review a Labor Department investigation of the wage relationship between anthracite and bituminous mining, but he refused to accept a general increase in bituminous wages merely because the miners were dissatisfied. Garfield viewed the unexpired Washington Agreement of October, 1917, which he had helped negotiate, as binding. Government, he believed, should formulate a unified wage policy, not deal with one industry at a time. He also refused to pass on a wage increase in higher prices. Despite appeals from his labor advisor and pleas from the mine leaders to the president, neither Garfield nor Wilson would budge.[1]

Following the armistice, plummeting demand and the prospect of excess labor as miners were mustered out of the army weakened the union's argument. Consumers switched to high-quality coals. Unemployment swept through many coal fields. By mid-winter, wage cuts in non-union fields left the union fighting just to maintain the Washington Agreement levels.[2]

At war's end, Garfield and other dollar-a-year men chomped at the bit—anxious to return to private life. "Now that the war conditions have abated," Secretary of War Newton Baker bluntly apprised Wilson, "all that is necessary for the industries of America to resume is that [wartime] restrictions should be relaxed and removed." Wilson told Congress on December 2, 1918, that he planned to remove economic controls. Several War Industries Board officials wanted them continued. But within two days of the armistice the agency removed a few restrictions, by the end of November it can-

1. UMW President Frank J. Hayes to Garfield, August 17, 1918; Garfield to Hayes, August 21, 1918; UMW district officers to Garfield, August 24, 1918; Garfield to Hayes, September 6, 1918; C. E. Baldwin, acting commissioner of labor statistics, to Garfield, September 24, 1918; Secretary of Labor William B. Wilson to Garfield, October 2, 1918; Garfield telegram to Hayes, October 24, 1918, "Report of the Conference Committee of the National Labor Adjustment Committees," October 14, 1918, all in Reel 280, Wilson Papers; Garfield to Hayes, September 6, 1918, Wilson Papers, Reel 293, Wilson Papers; Charles B. Fowler, *Collective Bargaining in the Bituminous Coal Industry* (New York: Prentice Hall, Inc., 1927), p. 83. James B. Neale to Garfield, August 18, 1918; Hayes, Lewis, and William Green to Garfield, October 21, 1918; Hayes, Lewis, and Green to Wilson, November 1, 1918; Garfield to Hayes, September 6, 1918; copy of telegram, J. P. White to Rembrandt Peale, November 1, 1918; Garfield to White, November 1, 1918, "Conference May 16th, 1918"; "Conference Held with Coal Miners, October 1, 1918," all in Records of Harry A. Garfield, General Correspondence File, Records of the United States Fuel Administration, RG 67, NA; *United Mine Workers Journal*, September 1, 1918, pp. 4–7; October 15, 1918, pp. 3–4; November 15, 1918, pp. 4–5.

2. Memorandum of February, 1918, by Bentley S. Warren to Garfield, Box 93, Garfield Papers; Frederick Lewis Allen, *Only Yesterday* (New York: Bantam Books, 1957; originally published by Harper and Brothers, 1931), pp. 10–13; U.S., Fuel Administration (hereafter cited as FA), *Final Report of the United States Fuel Administrator, 1917–1919* (Washington, D.C.: Government Printing Office, 1921), p. 164; U.S., Congress, Senate, Committee on Interstate Commerce, *Increased Price of Coal: Hearings Pursuant to S. Res. 126*, 66th Cong. 1st sess., 1920, pp. 46, 203; *United Mine Workers Journal*, December 15, 1918, p. 3; February 1, 1919, p. 6; March 1, 1919, p. 2.

celled priority orders, and a presidential order a month later dissolved the WIB. Having begun the dismantling of the Fuel Administration, Garfield returned to Williamstown by Christmas.[3]

Garfield longed, however, to maintain the benefits of wartime cooperation. He arranged a series of meetings among business, government, and labor February 11–14, 1919, to discuss ways to maintain the working relationships which the Fuel Administration had helped create. He asked the gathering of former staff, leading operators—primarily from the National Coal Association—and UMW spokesmen if they did not still have some unifying "common object." The extensive and valuable cost-of-living, price, and wage data collected by the government could, he thought, improve efficiency and stabilize employment and production.[4]

He sold them on the idea. "To return to the old unbalanced condition," proclaimed Central Coal & Coke's Harry N. Taylor, "would be almost criminal." Frank J. Hayes, the alcoholic and ineffective but still popular UMW president who succeeded John P. White when White joined the Fuel Administration, reminded the others that the Fuel Administration had eliminated instability and strikes. The industry, he feared, needed to be stabilized to preempt the collectivism of the rising Bolshevist movement. The operators and UMW officials initially balked at putting a governmental member on a postwar coal bureau, but Garfield persisted. The group finally backed the idea of a peacetime fuel agency.[5]

3. Baker to Wilson, November 30, 1918; Bernard M. Baruch, chairman, War Industries Board, to Baker, November 30, 1918; Baker to Wilson, June 17, 1918; Wilson to Baker, June 12, 1918, all in Box 8, Baker Papers; Robert D. Cuff, *The War Industries Board: Business-Government Relations during World War I* (Baltimore: Johns Hopkins University Press, 1973), pp. 243–61; John M. Blum, *Woodrow Wilson and the Politics of Morality* (Boston: Little, Brown, and Co., 1951), p. 141; William F. Willoughby, *Government Organization in War-Time and After* (New York: D. Appleton-Century Company, 1919), p. 114; Robert F. Himmelberg, "The War Industries Board and the Antitrust Question in November 1918," *Journal of American History* 52 (June, 1965): 59–74.

4. FA, *Final Report*, pp. 136–37, 161; the original copy of the conference is in Reel 293, Wilson Papers; Willoughby, pp. 18–20, 115.

5. FA, *Final Report*, pp. 138, 167; *United Mine Workers Journal*, March 1, 1919, pp. 3–6.

They urged Wilson to press Congress to create a fuel commission of three operators and three miners as advisors to a national fuel administrator who would license all coal industrialists. Licensed operators would receive exemption from prosecution of the antitrust laws as they met to establish a fair trade practices code to be "promulgated by such fuel administrator with the approval of the President." The administrator could review labor disputes when both parties agreed to mediation. He would also promulgate "rules and regulations as to the production, conservation, distribution, apportionment, storage and sale of coal."[6]

Garfield transformed this ad hoc committee's idea into a sweeping proposal to continue wartime cooperation among all American business, government, and labor. He envisioned a number of seven-man industrial commissions representing capital and labor in the country's basic industries. Headed by presidentially appointed directors, the commissions would have access to the wealth of facts accumulated by the permanent government departments and commissions and could recommend presidential or legislative action in one industry. The president could delegate administrative responsibilities to individual directors. Collectively they would form an industrial cabinet to advise him on general industrial problems. The directors, Garfield hoped, would force the spokesmen for labor and capital to reckon with the public interest in industrial decisions. He expected these directors to be replicas of himself, one suspects, and he saw his plan as analogous to the League of Nations. Garfield readied his proposal for Wilson's consideration when the president returned from Europe.[7]

At a final gathering of the Wednesday "industrial cabinet"—the original model for Garfield's proposal, which in-

6. FA, *Final Report*, p. 167.
7. "A Plan to Promote the Public Welfare by More Effective Cooperation between the Government of the United States and Industry," in Records of Harry A. Garfield, General Correspondence, RG 67, NA; also published in U.S., Congress, Senate, Committee on Manufactures, *Publication of Production and Profits in Coal, Hearings on S. 4828*, 66th Cong., 3rd sess., 1921, pp. 1259ff. (hereafter cited as *Publication of Production and Profits*); Willoughby, p. 302; Cuff, "Harry Garfield, the Fuel Administration, and the Search for a Cooperative Order during World War I," *American Quarterly* 30 (Spring, 1978): 41–52.

cluded the administrative heads of the various war agencies—Garfield eagerly expounded his plan. Wilson asked questions, but doubted that the plan would pass Congress at the present session. Garfield suggested that Wilson institute some variation of the proposal by executive order. Knowing that such an order would enrage those weary of federal controls, Wilson refused. He preferred, he said, using the heads of permanent governmental agencies and commissions to creating new directors. Garfield consulted with the other war administrators and began to reformulate the plan.[8]

The new plan, unveiled in April, 1919, proposed that each basic industry structure a commission of capital and labor to advise particular department secretaries. Bituminous coal, anthracite coal, and petroleum would each have a commission under the direction of the secretary of the interior. Steel, copper, brick, cement, and public utilities would report to the secretary of commerce. The cabinet heads involved and the industrial administrators would form a new industrial cabinet. The industrial administrators would, Garfield hoped, use government statistics to settle controversies in the public interest. On April 9 Garfield sent the new plan to Wilson in Paris.[9]

Wilson was too concerned with negotiating a peace treaty in Versailles, and business and labor were too eager to break free from wartime governmental interference to support Garfield's plan. Like the U.S. Chamber of Commerce's Reconstruction Commision, the WIB's attempts at antitrust revision, the Department of Commerce's Industrial Board, and other postarmistice proposals to continue the cooperative wartime programs, Garfield's visionary proposal deviated too much from accepted peacetime roles for government, business, and labor. The enthusiasm expressed at the mid-

8. Harry A. Garfield, untitled diary account of the meeting, February 26, 1919, in Subject File, U.S. Fuel Administration, Garfield Papers; Garfield testimony in *Publication of Production and Profits*, p. 1248; Garfield to Wilson, March 3, 1919, Reel 293, Wilson Papers.

9. "Further Elaboration of the Plan," Box 93, Garfield Papers. Garfield to Secretary of the Navy Josephus Daniels, March 14, 1919; Garfield to Wilson, March 27, 1919; Garfield memorandum to the president (no date), all in Records of the Fuel Administration, General Correspondence, RG 67, NA; *Publication of Production and Profits*, pp. 1260, 1300–14.

February meetings evaporated. Spokesmen for the NCA and the UMW balked at offering public endorsement. When operator and union disagreements led to the 1919 strike, Garfield bitterly mourned the rejection of his proposal.[10]

Staged against a backdrop of apparent worldwide revolution, the coal strike of October, 1919, further eroded the wartime spirit of business-governmental cooperation. Soviet-style revolutions erupted in Eastern Europe, radical leaders splintered the American Socialist Party with calls for "an American Soviet," and Karl Radek, executive secretary of the Third International, which met in March, 1919, boasted that the funds sent for the Berlin Spartacist Revolution "were as nothing compared to the funds transmitted to New York for the purpose of spreading bolshevism in the United States." To a nation suffering from the economic and psychic dislocation of war—including antiwar agitation and strikes by American socialists and the International Workers of the World—the call by radical unionists in February, 1919, for a general strike in Seattle confirmed the public's worst fears. The UMW policy committee endorsed nationalization of the mines. If this was not Bolshevism, what was? Two waves of coordinated bombings rocked the homes of various public figures in the spring, and the nation plunged into the hysteria of the Red Scare.[11]

The wave of strikes—particularly those in steel and coal of the fall—seemed just the opening volleys of revolution. Em-

10. George Soule, *Prosperity Decade: From War to Depression, 1917–1919,* (New York: Holt, Rinehart and Winston, 1968; originally published by Harper and Row, 1947), pp. 62–63; George Peek, *History of the Industrial Board of the Department of Commerce* (privately published, 1919), p. 49; James Weinstein, *The Corporate Ideal in the Liberal State: 1900–1918* (Boston: Beacon Press, 1968), pp. 231–33; Willoughby, pp. 116–20. National Coal Association, "National Coal Association Referendum No. 1: Proposed Plan of Cooperation"; Garfield to Frank J. Hayes, April 16, 1919; Garfield to J. D. A. Morrow, March 18, April 19, 1919; Morrow to Garfield, April 17, 1919, all in Records of Harry A. Garfield, General Correspondence, Plan, RG 67, NA; FA, *Final Report,* pp. 37–38.

11. Robert K. Murray, *Red Scare: A Study in National Hysteria, 1919–1920* (New York: McGraw-Hill Book Company, 1964; originally published by University of Minnesota Press, 1955), chapters 1–5; Stanley Coben, "A Study in Nativism: The American Red Scare of 1919–20," *Political Science Quarterly* 79 (March, 1964): 52–75; Selig Perlman and Philip Taft, *History of Labor in the United States, 1896–1932* (New York: Macmillan Company, 1935), chapter 36; United Mine Workers of America, Nationalization Research Committee, *Compulsory Information in Coal* (UMWA, 1922), pp. 16–19; *United Mine Workers Journal,* April 1,

ployer and public attitudes hardened. Wildcat strikes erupted in protest to layoffs and wage cuts. Illinois operators imposed the dollar-a-day wartime penalty clause on strikers. Because price restrictions on coal had been suspended in February, the UMW argued that although no treaty had been ratified, the war was over, and that the Washington Agreement had expired. At the September UMW convention, the leadership demanded 60 percent increases in wages, the six-hour day and five-day week, an end to penalty fines, and a nationwide contract with no customary sectional settlements. If a satisfactory agreement was not secured before November 1, 1919, the miners would strike.[12] Thirty-seven-year-old UMW Vice-President John L. Lewis took over the acting UMW presidency from the ailing Frank Hayes. The stout, blue-eyed Lewis had risen to prominence in a series of swift steps. Better than anyone else around, he commanded the statistical data of the coal industry. Powerfully built, conservative in dress, but with a stony face and a head of wavy auburn hair, Lewis would soon become the most significant single individual in the industry. On the back of a 1917 photograph he had cryptically written, "Jesse James in disguise." Would he become the highwayman of the industry—or perhaps of the union?[13]

Power also shifted in the nation's capital. Exhausted from a demanding fall tour of the West in defense of his League of Nations, President Wilson suffered two strokes that paralyzed his entire left side. He denied the existence of his disability and made feeble attempts to hide its importance from his cabinet and the public, but he became physically

1919, p. 3; Lewis Lorwin, *The American Federation of Labor* (Washington: The Brookings Institution, 1933), p. 199; *New York Times*, January 1, May 17, 1923; John Brophy, *A Miner's Life: An Autobiography*, ed. John O. P. Hall (Madison, Wis.: University of Wisconsin Press, 1964), pp. 151–57, 170–72; John Brophy, "Elements of a Progressive Union Policy," in J. B. S. Hardman, *American Labor Dynamics* (New York: Harcourt Brace Co., 1928), p. 191; Glen Lawhon Parker, *The Coal Industry: A Study in Social Control* (Washington, D.C.: American Council on Public Affairs, 1940), p. 94.

12. Fowler, pp. 84–85; Philip Taft, *Organized Labor in American History* (New York: Harper and Row, 1964), p. 350; *Publication of Production and Profits*, p. 623; *United Mine Workers Journal*, April 1, 1919, p. 3.

13. Melvyn Dubofsky and Warren Van Tine, *John L. Lewis* (New York: Quadrangle/The New York Times Book Co., 1977), chapter 2 and p. 153; Saul Alinsky, *John L. Lewis: An Unauthorized Biography* (New York: G. P. Putnam's Sons, 1949), pp. 26–27, 31–36; *United Mine Workers Journal*, October 1, 1919, p. 7; November 1, 1919, p. 7.

unable to govern. Effectively shielded by his wife, Mrs. Edith Bolling Galt Wilson, and Admiral Cary Grayson, his personal physician, from October, 1919, on, Wilson's presidential powers slipped into the hands of those willing to fill the void.[14]

The deadline approached. Secretary of Labor William B. Wilson brought pressure on both union and management for a conference, but Lewis issued a strike call in advance of the meeting. The operators offered only a moderate wage increase. Operator spokesman Thomas T. Brewster accused the Kremlin of ordering the strike and financing it with Bolshevik money. A. Mitchell Palmer, the "fighting Quaker" who had become the attorney general, moved onto center stage.[15]

The cabinet met without Wilson on October 25 and appointed Palmer, Secretary of Labor William Wilson, and Interior Secretary Franklin K. Lane to redraft a message Presidential Secretary Joseph Tumulty had written. Signed by the president, the statement damned Lewis's strike order as "a grave moral and legal wrong." Under pressure from Palmer and other cabinet officials, the president restored by executive order the Fuel Administration's maximum coal prices under the Lever Act. The reactivated law prohibited any conspiracy to limit supply or distribution of coal. Palmer then secured an injunction against the union's strike. Lewis refused to rescind the strike order, and the coal miners of America walked out of the pits. Amid a flurry of press accounts labeling the strike a Bolshevik action, Palmer got

14. Wilson saw a few visitors during the winter of the coal strike but did not meet with the cabinet until April 14, 1920, and then only as an invalid, propped up in a White House study near his bedroom. After an hour with the cabinet in which Wilson urged the attorney general not to "let the country see Red," he found himself failing to concentrate on the discussion of nationalization of the railroads, and his wife suggested that the meeting end. Gene Smith, *When the Cheering Stopped: The Last Years of Woodrow Wilson* (New York: Bantam Books, Inc., 1965; originally published by William Morrow & Company, 1964), chapters 6–10, especially pp. 149–50; Robert K. Murray, *The Harding Era: Warren G. Harding and His Administration* (Minneapolis: University of Minnesota Press, 1969), pp. 76–81; Edwin A. Weinstein, "Woodrow Wilson's Neurological Illness," *Journal of American History* 57 (September, 1970): 324–51.

15. A strike by railroad shopmen in August, 1919, disrupted coal transportation and reactivated the Fuel Administration. But because the agency lacked appropriations, distribution of coal was handled through the office of director general of the railroads. A. G. Gutheim, "The Transportation Problem in the Bituminous Coal Industry," *American Economic Review* 11 (March, 1921): 102; Lorwin, pp. 184–85; Murray, *Red Scare*, pp. 154–57; Soule, p. 195; Fowler, pp. 86–88.

another injunction on November 8 ordering the UMW leadership to cancel the strike. The UMW executive committee went into a grueling, seventeen-hour session on November 10–11. Lewis finally emerged to concede.[16]

But the striking miners stayed out. Delegated by the cabinet, Secretary of Labor Wilson assembled a joint conference of operators and union leaders for November 14. Reactivated as fuel administrator, Garfield joined them. His statistics convinced him that a 14 percent increase sufficed to bring the miners even with the wartime rise in the cost of living. He assumed the role of spokesman in the public interest as described in his industrial plan. Any higher levels, he maintained, would require an intolerable increase in prices. What had disrupted their earlier support for his plan for industrial harmony, he asked? He pleaded the statistical case for trying to keep the wage settlement from pushing the cost of living still higher. "The problem," William Green, UMW secretary-treasurer, argued tartly, "can not be solved as a college professor would work out a problem in geometry, algebra, or theoretical philosophy."[17] Garfield's olympian presentations also angered Secretary Wilson, who thought the miners deserved an increase nearer 31 percent to catch up with living costs. At cabinet sessions he attacked the fuel administrator's meddling. Garfield's insistence on holding prices firm, he felt, nullified the negotiation process and undermined him. He threatened to resign.[18]

Garfield believed in the dominance of an impartial public interest spokesman. The president, he thought, supported

16. Smith, pp. 118–19, chapters 7–10. Robert Murray has written that the cabinet did not meet prior to Palmer's seeking the injunction. He based his conclusion on Secretary of State Lansing's *Desk Diaries* (*Red Scare*, note 11, p. 157, and pp. 310–11). Secretary of the Navy Daniels' cabinet diaries provided the evidence for the meeting I have described (Josephus Daniels, *The Cabinet Diaries of Josephus Daniels*, ed. E. David Cronon [Lincoln: University of Nebraska Press, 1963], October 25, 1919, p. 452; see also pp. 453–56). Daniels noted that Wilson "expressed deep regret that Palmer had asked an injunction and thought in present situation no Commission should be appointed . . ." (Daniels, *Cabinet Diaries*, p. 456); Wilson to Henry P. Robinson, December 19, 1919, Reel 294, Wilson Papers.

17. FA, *Final Report*, pp. 17–18; Green quoted in *United Mine Workers Journal*, December 1, 1919, pp. 3–5, 12.

18. "Memorandum of the Secretary of Labor Relative to His Position on the Coal Strike" (no date), in Box 129, Garfield Papers; Daniels, *Cabinet Diaries*, November 25, 1919, p. 464.

him. Other cabinet officials endorsed Garfield's concepts in a general way, but some realized apparently that the settlement would have to be mediated, not dictated by Garfield or anyone else. The operators believed that they would have to give more than 14 percent on wages and were willing to pass on increased costs in higher prices, precisely what Garfield hoped to avoid.

On December 6, 1920, after a federal judge cited eighty-four union officials for contempt of court, Lewis arranged through a mutual friend to meet with Palmer. The day before, cabinet members had agreed to have Palmer give both Garfield's and Secretary Wilson's positions to the president to consider. Palmer brought from the president a statement which reviewed the arguments of the two men and said that the president would appoint a government tribunal as suggested by Garfield if the miners accepted the 14 percent increase and returned to work. Then the president's tribunal would review both wages and prices. Failure to return to work under these conditions, the statement said, would be unjustifiable. By continuing to defy the injunction, the miners were testing the ability of government to enforce the law. Palmer outlined the government's position to Lewis and William Green that afternoon without making it clear that President Wilson had intervened. The labor leaders protested that they had already agreed to accept the 31 percent offer made by the secretary of labor. No deal.[19]

Palmer reported to Garfield early that evening on the afternoon meeting with the union leaders, but he said nothing about Wilson's commission with full power to renegotiate both wages and prices. When Palmer met with Lewis and Green again at 10 P.M., he offered to have the president establish a three-man commission with one man acceptable to the miners, one for the operators, and one as public representative. Putting the public member in a minority on such a commission violated all of the principles Garfield had been fighting for. But it settled the strike. Palmer withheld

19. Daniels, *Cabinet Diaries*, December 5, 1919, p. 467; typescript copy of Wilson's statement, Box 129, Garfield Papers; *United Mine Workers Journal*, December 15, 1919, p. 3.

the president's statement from the press until the union had time to win district representatives' approval to end the strike. Only when an Associated Press reporter telephoned him at 1:30 A.M. seeking details of the settlement did Garfield learn that his position had been undercut. Dumbfounded, he met with the attorney general the next morning. The new commission could accept Garfield's position and refuse a further increase, Palmer argued, but his offer to Lewis and Green had already been made.[20]

Garfield wrote President Wilson to get his support for the principle of maintaining the public interest by keeping the commission advisory only and reserving final judgment to a public official. Admiral Grayson telephoned Garfield the next day that the president had written "Approved" across Garfield's letter and was returning it via Secretary Tumulty. Garfield asked Grayson that afternoon if he, Garfield, could telegraph the letter to Palmer, then en route to Indianapolis with the mine leaders. Grayson told him "that the president left the matter entirely to my [Garfield's] discretion; I could use the letter when and how I pleased; that he saw the position I had been put in and relied upon me to use my best judgment." Garfield felt that his stand had been endorsed by Wilson and thus would prevail.[21]

Following a cabinet meeting on the ninth, however, Mrs. Wilson sent Garfield a handwritten note that President Wilson wished to withhold "his approval of the letter you sent him on December 7th. He fears that any qualification, at this stage, of the terms of his letter of Saturday suggesting the creation of a Commission would bring about a serious setback in the whole settlement, and thinks that the best course is to leave the Commission perfectly free. . . ." The fuel administrator inferred that Tumulty had kept Wilson's approval of Garfield's position from the cabinet meeting that morning and had worked to change the president's earlier

20. Garfield to Robert Lansing, January 13, 1920, Box 93, Garfield Papers. This letter contains copies of correspondence between Garfield, William Green, Wilson, and Palmer; *United Mine Workers Journal*, December 15, 1919, p. 6.

21. Copies of these messages and much of the background material are in Garfield's diary (typewritten), December 9, 10, 11, 1919, Box 127, Garfield Papers; see also Garfield to Lansing, January 13, 1920, Box 9, Garfield Papers.

decision. The cabinet that morning, Garfield noted, did not include his supporters, in any case. Lansing counseled Garfield to wait for the commission's findings, but Garfield asked Wilson on the eleventh to accept his resignation. The president reluctantly accepted it.[22]

Wilson appointed a commission which finally allowed a total increase of 27 percent in March, 1920, bringing miners' wages to historic highs. Garfield had battled to keep alive the spirit of business-management cooperation developed during the war but had lost. The president's illnes, he maintained long after, had allowed cabinet members to manipulate the chief executive. If Wilson had accepted his plan of industrial commissions with power vested in the public members, Garfield argued, the strike of 1919, considered by the *United Mine Workers Journal* to be "the greatest and most extensive industrial struggle ever staged in America," could have been prevented.[23]

In 1920 events confirmed Garfield's worst fears. The wage settlement raised costs. The long strike had reduced consumers' stocks; a 1920 wildcat strike by railway switchmen further hampered production. Unprecedented demand for export coal hit in a time of short supply. Without Fuel Administration price controls, 1920 coal prices skyrocketed. Average spot prices in March were $2.58; in April they jumped to $3.85. In May they leapt to $4.59, in June to $7.18, and in July to $8.24. The coal panic of 1920 peaked in

22. Edith Bolling Wilson to Garfield, December 9, 1919; copy, Garfield to Wilson, confidential, December 11, 1919; copy, Garfield to Wilson, December 11, 1919; Wilson to Garfield, December 13, 1919, in Box 127, Garfield Papers; Daniels, *Cabinet Diaries,* December 9, 1919, p. 468. The *United Mine Workers Journal* editorialized that Garfield's resignation was "one of the most pleasing results of the settlement" (December 15, 1919, p. 16).

23. Daniels, *Cabinet Diaries,* December 9, 1919, p. 468; C. L. Sulzberger, *Sit Down with John L. Lewis* (New York: Random House, Inc., 1938), p. 46; *Publication of Production and Profits,* pp. 724–25. Letters praising Garfield are in Subject File: Resignation, Garfield Papers; *New York Times,* January 26, March 17, 24, 1920. A stenographic report of the meetings of the Bituminous Coal Commission and the commission's reports are in Reel 294, Wilson Papers. Wilson to Henry P. Robinson, chairman of the commission, December 19, 1919, Reel 294, Wilson Papers; FA, *Final Report,* p. 38; Carnes, p. 24; Senate Interstate Commerce Committee, *Increased Price of Coal,* pp. 522–25, 542–54, 650–57, 688–99; *United Mine Workers Journal,* December 15, 1919, p. 6; January 1, 1920, pp. 3–5, 8; December 1, 1920, p. 4.

August, when spot prices averaged $9.51, more than twice the highest average for spot prices in the prewar boom (see Fig. 1, p. 20). In the quarter prior to the commission settlement, reports showed Ohio operators earning 5 percent on investment; that summer they earned 52 percent in June and 59 percent in July and August.[24]

Fostered by the political economy of war, Garfield's drive for business, government, and labor cooperation collapsed with the peace. Government intervention became anathema. Palmer's injunction embittered the unionists. Garfield's attempt to dictate a settlement through a commission angered the operators. The wartime sense of purpose and patriotism evaporated. Since it got price increases, management accepted higher wages. But neither management nor labor wished to subordinate its private interest to a statistically oriented commission. That fall an electorate weary of Wilsonian war controls gave the largest plurality in a century to Warren G. Harding, a middle-of-the-road politician from Ohio. In his inaugural address Harding proclaimed that he stood "for the omission of unnecessary interference of government with business, for an end of government's experiment in business, and for more efficient business in government administration."[25]

Harding, High Prices, and Herbert Hoover

Despite Harding's conservative, probusiness, and antireform bias, the 1920's can no longer be seen as a "reactionary and barren interlude between progressivism and the

24. U.S., Federal Trade Commission (henceforth cited as FTC), *Preliminary Report on Investment and Profit in Soft-Coal Mining* (Washington, D.C.: Government Printing Office, 1922), p. 8; F. G. Tyron, "Effect of Competitive Conditions on Labor Relations," *Annals of the American Academy* 111 (June, 1924): 90; *Publication of Production and Profits*, pp. 2051–55, 2084; U.S., National Recovery Administration (henceforth cited as NRA), "Economic Survey of the Bituminous Coal Industry under Free Competition and Code Regulation" (mimeographed), by F. E. Berquist, et al., Work Materials No. 69, 2 vols. (Washington, D.C., 1936), vol 1, p. 54.

25. *The Black Diamond*, December 27, 1918, p. 600; Himmelberg, pp. 59–74; Cuff, *War Industries Board*, p. 248; *United Mine Workers Journal*, December 1, 1920, p. 3. *Inaugural Address of President Warren G. Harding, Delivered Before the Senate of the United States*, 67th Cong., special sess., S. Doc. 1, 1921; Murray, *Harding Era*, p. 111.

New Deal." Many progressive impulses such as the War
Finance Corporation, the drive for conservation, and the
thrust for farm price parity carried through the decade. Secretary of Commerce Herbert Hoover attempted to stabilize
bituminous coal, stimulated trade associations, and sponsored Department of Commerce conferences on efficiency
and standardization. General Electric's Owen D. Young and
other corporate managers spoke of the need for business
responsibility to the consumer and the public. Such men
developed welfare capitalism and industrial democracy to
create a new form of American corporatism in many industries. "The more or less unconscious and unplanned activities of businessmen," Walter Lippmann wrote, "are for once
more novel, more daring, and in general more revolutionary
than the theories of the progressives." But could Hoover
and others reactivate the drive for stabilization in the highly
competitive bituminous coal industry?[26]

In April, 1920, when the National Coal Association won a
court decision prohibiting the FTC from acquiring the kind
of information it had collected during the war, the stabilizers
got part of the answer. In *Maynard Coal Company* v. *Federal
Trade Commission*, the District of Columbia Supreme Court
held that the commission's demands for detailed price and
cost information exceeded the commerce power of the Constitution. The court ruled that "no such visitorial power as
that claimed by the commission in the instant case has been
vested in Congress by the Constitution, nor could Congress
delegate such power to the commission.[27]

26. Ellis W. Hawley, "Secretary Hoover and the Bituminous Coal Problem,
1921–1928," *Business History Review* 42 (Autumn, 1968): 247; Morrell Heald,
"Business Thought in the Twenties," *American Quarterly* 13 (Summer, 1961):
126–36; Lippmann quoted in William E. Leuchtenburg, *The Perils of Prosperity,
1914–1932* (Chicago: University of Chicago Press, 1958), p. 202; Paul Glad, "Progressives and the Business Culture," *Journal of American History* 53 (June, 1966):
75–81; Paul Carter, *The Twenties in America* (New York: Thomas Y. Crowell
Company, 1968), p. 37, 42–43; J. Joseph Huthmacher and Warren I. Sussman, eds.,
Herbert Hoover and the Crisis of American Capitalism (Cambridge, Mass.: Schenck-
man Publishing Company, 1973); Ellis W. Hawley, "Herbert Hoover, the Commerce Secretariat, and the Vision of an 'Associative State,' 1921–1928," *Journal of
American History* 61 (June, 1974): 116–40; Murray, *Harding Era*, pp. 195–98 and
passim.

27. *Maynard Coal Company* v. *Federal Trade Commission*, quoted in Parker, p.
90; *Publication of Production and Profits*, pp. 377–80.

Undaunted, Senator William M. Calder, a conservative Republican from New York, charged that the bituminous industry had earned excessive profits in 1920, and he introduced a bill to empower the Department of Commerce to collect statistical information on production and profits. In the lengthy hearings senators found that the UMW, recoiling from the government's use of the Lever Act against it, opposed even a section which directed payment of a fair wage. Speaking for his members, William Green said that the union was like "a burned child; they always fear fire." Strong opposition by the NCA and consumer apathy doomed the measure.[28]

Two other congressional proposals, one to have the ICC lower freight rates in summer months to encourage year-round operation of the mines and another to create a coal commission to gather and publish statistical data, also fell to defeat. The ICC supported the first measure, and Secretary of Commerce Hoover supported the commission measure, as did some enlightened northern operators. But the NCA's annual convention stood "unalterably opposed to the enactment of any legislation imposing additional regulation upon commerce and industry, and is especially opposed to legislation which singles out any one industry for regulation by special commission." Although the association was losing its membership in the unionized fields, Hoover was wrong in blaming the failure of the second measure on "paid lobbyists from some minor branches of the industry." Few operators sought assistance from government in 1920.[29]

Operators and miners were particularly self-reliant in the rich coal fields in the hills of West Virginia, where disputes often were settled with guns. A May, 1920, strike over the discharge of union workers in Mingo County erupted in a gunfight between union sympathizer Sid Hatfield, chief of police in Matewan, and Albert and Lee Felts of the Baldwin-

28. Green testimony in *Publication of Production and Profits,* pp. 1224–25; Parker, p. 90; William Kenyon, "A Code of Industrial Law," *Annals of the American Academy* 111 (January, 1924): 306.

29. *Publication of Production and Profits,* pp. 881–914, 945, 972, 1052–83, 1102. The minutes of the NCA meeting of May 27, 1920, are quoted on p. 2019. Hoover quoted by Hawley, "Hoover and the Coal Problem," p. 255.

Felts detective agency. Six men died in the exchange of bullets. Civil war ensued. Further gunfights killed six more. By midwinter the conflict had produced sixteen murders. To revenge an attack by deputy sheriffs on striking miners in a tent colony, a miners' army of over four thousand planned an armed invasion of non-union territory. Reinforced by additional recruits arriving on a commandeered train, the union army tramped toward an opposition force of some two thousand. The governor appealed to the president for help.[30]

Both sides refused to heed a presidential order to disperse. Harding dispatched Army General Bandholtz and two thousand men of the Nineteenth Infantry into Logan County. With air support from Billy Mitchell, the infantry ended the West Virginia war. Six hundred union soldiers surrendered to the federal troops and were disarmed. Although more than five hundred men were indicted for various offenses from treason to conspiracy, the charges against them were later dropped.[31] The charge of chaotic labor-management relations in the bituminous industry, however, could not be dismissed.

Following an investigation into the West Virginia mine fields by the Senate Committee on Education and Labor, the chairman, William S. Kenyon, Iowa Republican, launched a drive for the enactment of an industrial code. Impressed by the National War Labor Board's 1918 industrial code, Kenyon regretted that Wilson's two industrial conferences at the end of the war had not produced a "Code of principles for the determination of industrial disputes." Careful to distinguish his proposal from the rigid arbitration board of the Kansas Court of Industrial Relations, he proposed a code like the one

30. Alinsky, pp. 40–41; Jerold Auerbach, *Labor and Liberty: The La Follette Committee and the New Deal* (Indianapolis and New York: Bobbs-Merrill Company, 1966), pp. 22–23; U.S., Congress, Senate, Committee on Mines and Mining, *To Create a Bituminous Coal Commission: Hearings before a Subcommittee,* 72d Cong., 1st sess., 1932, pp. 1126–27, 1324, (hereafter cited as *To Create a Commission*); Sidney Lens, *The Labor Wars: From the Molly Maguires to the Sitdowns* (Garden City, N.Y.: Anchor Books, 1974; originally published by Doubleday & Company, 1973), pp. 266–70; *United Mine Workers Journal,* June 1, 1921, p. 6; August 15, 1921, p. 3.

31. Perlman and Taft, pp. 479–81; McAlister Coleman, *Men and Coal* (New York: Farrar & Rinehart, 1943), pp. 99–104; Taft, pp. 353–54; *United Mine Workers Journal,* June 15, 1921, p. 14; September 1, 1921, p. 4; September 15, 1921, pp. 3–4, 8.

in the 1920 Transportation Act creating the Railway Labor Board. "This foundation in the law [Kenyon's code] dealing with the modern conditions of social and economic life," President Harding hoped, "would hasten the building of the temple of peace in industry which a rejoicing nation would acclaim."[32]

Kenyon's code proposal prefigured the industrial principles developed under the National Industrial Recovery Act and the later National Labor Relations Board, but like the Garfield proposal, it attracted few supporters in the 1920's. Kenyon urged Congress to declare coal to be a public utility in which "the public interest is predominant." He wanted to sanction collective bargaining rights (including the right not to join a union), establish the six-day week and the eight-hour day with a basic wage and overtime pay, abolish child labor, get equal pay for women working in industry, protect investment, and outlaw strikes until a hearing determining the facts could be held. But Kenyon's fellow committeemen withheld support.[33] For a decade following the issuance of his report in 1922, the idea of a coal code was just an idea.

A mining engineer by profession, Commerce Secretary Herbert Hoover developed his own plan for stabilizing the bituminous industry and providing more continuous employment in 1922. To attack the excess capacity of the industry he would penalize mines that worked less than 240 days a year and use the penalty funds to pay compensation to less efficient mines that shut down and placed themselves in a pool to be tapped only in times of peak demand. To improve marketing, Hoover wanted cooperative district associations to sell coal and "divide the receipts from sales to the individual operators upon the basis of their proportionate production of each grade or quality." Competition between districts and with the independent operators would maintain free prices. The cooperatives should be exempt from the

---

32. Harding quoted in U.S., Congress, Senate, *West Virginia Coal Fields*, Report No. 457, 67th Cong., 2d sess., 1922, p. 9; Kenyon, pp. 307–12; *United Mine Workers Journal*, July 15, 1921, p. 14; February 15, 1922, pp. 3–6.
33. Hawley, "Hoover and the Coal Problem," p. 257; Senate, *West Virginia Coal Fields*, pp. 1–28; *United Mine Workers Journal*, February 15, 1922, p. 6.

antitrust laws, Hoover believed, if they filed with the FTC and as long as there were "no fixing of prices or profits."[34]

The elimination of excess mine capacity and gradual reduction in the bituminous work force, Hoover argued, combined with better marketing and transportation, would "secure regular operation," permit larger credit sales, reduce cross-hauls, speed up traffic, save from $0.50 to $1.00 a ton, produce stable prices, expand foreign trade, and minimize labor troubles. Hoover expected "demagogues" to cry "coal trust," and in an inquiry to Harry Garfield, Hoover conceded that "to satisfy the public" he might have to add some kind of price or profit controls. But these, he thought, tended to "break down initiative." As Hoover consulted Garfield and others, however, the nation suffered through the biggest strike in the industry's history.[35]

The hothouse economy of the war cooled. Government expenditures declined, the railroads retrenched, the Federal Reserve Banks raised interest rates, business slowed purchases, and the recession of 1921 began. Wartime coal production of nearly 600 million tons in 1918 fell to 415 million tons in 1921. The unionized mine fields in southern Ohio lost work time as non-union fields to the south and east slashed wages. Half the miners worked only half-time. Impressed by Judge Elbert Gary's maintenance of the open shop in steel, and with Harding's acquiescence, and with 60 million tons of above-ground coal, the operators sought to destroy the union.[36]

34. Hoover, "Plan to Secure Continuous Employment and Greater Stability in the Bituminous Coal Industry" (draft 1, May 9, 1922), Commerce Papers, Coal, E. E. Hunt Coal Plans, Box COF 103, Hoover Papers; "Summary of Address of Herbert Hoover, September 12, 1922, 'The Public and the Coal Industry,' " Commerce Papers, Coal, Mr. Hoover's Plan, Box COF 106, Hoover Papers; Hawley, "Hoover and the Coal Problem," pp. 259–62; Robert F. Himmelberg, *The Origins of the National Recovery Administration* (New York: Fordham University Press, 1976), pp. 66–72.

35. Hoover, "Plan to Secure Continuous Employment"; Hawley, "Hoover and the Coal Problem," p. 261; Hoover to Garfield, May 2, 1922; Garfield telegram to Hoover, May 6, 1922, all in Commerce Papers, Coal, Mr. Hoover's Plan, Box COF 106, Hoover Papers. Garfield naturally suggested structuring the organization along the lines he had advocated in 1918–19.

36. Soule, pp. 96–115; M. B. Hammond, "The Coal Commission Reports and the Coal Situation," *Quarterly Journal of Economics* 38 (1924): 559; Brophy, "Elements of a Progressive Union Policy," p. 187; Tyron, p. 91; Fowler, pp. 94–97; John M. Clark, *The Costs of the World War to the American People* (New Haven: Yale University Press, 1931), pp. 58–59; Murray, *Harding Era*, pp. 84–86.

Harding failed to reconcile the two sides. Southern Ohio and western Pennsylvania—two of the crucial divisions of the Central Competitive Field nearest the non-union Appalachian fields—refused to bargain with the UMW. Discussions collapsed. The UMW refused to negotiate by districts, and on April 1, 1922, when the current contract expired, 610,000 miners—both anthracite and bituminous—struck.

In Williamson County, Illinois, over the protests of the Illinois National Guard, the Southern Illinois Coal Company employed members of the Chicago Steamshovel Men's Union to strip-mine coal. In attempting to protest this scabbing, two UMW men were killed in machine-gun fire. Private police guarded the steam shovelers. Striking miners armed and advanced on the stockade erected around the company's works in Herrin. The shovelers waved the white flag of truce and surrendered, but the aroused miners violently killed eighteen of the disarmed scabs. Twenty-one persons died in the Herrin massacre.[37]

Coupled with a threatened strike by railroad shopmen and an increasing danger of a serious coal shortage, the shock of the Herrin massacre pushed a reluctant and overwrought Harding to intervene. In mid-May he picked Herbert Hoover, who was beginning to distrust the "swine" among the operators, as head of a Coal Distribution Committee. Hoover got some operators to lower prices. Harding brought representatives of both sides to Washington in May. Beginning on July 1, he launched a "final effort." After these discussions collapsed, the president on July 10 appealed to both sides to resume work at the old terms, to allow an eleven-man coal commission to arbitrate changes in wages and working conditions, and to support an exhaustive investigation of the industry. No takers. Although Harding felt that the union sought "national dictation" of wage agreements, he realized that some operators were "warring on the unions." Stung by the mine unionists' refusal to return to

37. Perlman and Taft, pp. 482–84; Fowler, pp. 93, 98; Hammond, "Coal Commission Reports," p. 544; *New York Times*, June 22, 1922; *United Mine Workers Journal*, January 15, 1922, pp. 3–4, 6; March 1, 1922, p. 10; April 1, 1922, pp. 3–5; July 1, 1922, pp. 3–7.

work, Harding requested the governors of twenty-eight states to protect properties of working mines. Under a proclamation of martial law, state troopers and national guardsmen protected volunteers who mined 1,754 tons of coal at a cost of $37.35 per ton, including the expense of security.[38]

A large segment of the bituminous community cooperated with the voluntary program of priority movement and lowered prices, but the shortages brought by the 1922 strike led Congress to demand direct controls. With scarcely any opposition, a bill supported by Harding and Secretary Hoover created in late September a federal fuel distributor to determine where shortages existed, to evaluate the "reasonableness" of prices, and to establish priorities among consumers. Although the bill was modeled in part on the Lever Act, Hoover made it clear that the fuel distributor was not another fuel administrator. This distributor would work primarily through the Interstate Commerce Commission to insure priority movement.[39]

In appointing Conrad E. Spens, vice-president of the Chicago, Burlington, and Quincy Railroad, as federal fuel distributor, Harding directed him to solve the coal shortage "by cooperation rather than by regulation." Julius Barnes, president of the United States Chamber of Commerce, pleaded that consumers reduce unnecessary purchases. James D. A. Morrow feared that "rigid application" of the ICC priority orders might put the "life and death" of the coal industry into the ICC's hands. He was wrong. But many miners returned to work and eradicated most of the shortages by Christmas of 1922. The ICC canceled priority

38. *To Create a Commission*, pp. 534, 536–37; Hammond, "Coal Commission Reports," pp. 545; Fowler, pp. 98–100; *New York Times*, May 17, June 1, July 1, 11, 29, August 19, 1922; *United Mine Workers Journal*, July 1, 1922, pp. 16, 22; Andrew Sinclair, *The Available Man: The Life Behind the Masks of Warren G. Harding* (New York: Quadrangle Books, 1969; originally published by the Macmillan Company, 1965), p. 256. Harding became so deeply involved in the labor-management relations of the coal industry during the summer of 1922 that he forsook golf. Murray, *Harding Era*, pp. 168, 262–63, 388–89; Hawley, "Hoover and the Coal Problem," p. 258.

39. Herbert Hoover, *The Memoirs of Herbert Hoover*, 2 vols. (New York: The Macmillan Company, 1952), vol. 2, p. 71; *Statutes at Large*, vol. 42, pt. 1, pp. 1025–28; U.S., Congress, House, Committee on Interstate and Foreign Commerce, *Coal: Federal Fuel Distributor*, Hearing on H.R. 12472, 67th Cong., 2d sess., 1922, pp. 2–39; *New York Times*, July 29, 30, August 19, September 1, 16, 23, 1922.

rulings to take effect in mid-December, and in early 1923, Harding scrapped the program.[40]

John L. Lewis's overly optimistic call the summer before for a wage conference in Cleveland brought operators representing only 20 percent of the tonnage of the old Central Competitive Field. Both sides, however, had reached certain economic limits. Lewis had expended close to four million dollars in relief funds; some operators had not mined coal for five months. The prospect of high prices, fear of an injunction, and pressure from Treasury Secretary Andrew Mellon brought the operators and unions to the table. Worked out in September and October and copied elsewhere, the Cleveland Agreement maintained the old wage terms until April 1, 1923, outlawed the automatic penalty clause, and provided for renegotiation following an investigation by a U.S. Coal Commission created by Congress in September. The Cleveland formula settled the strike for most of the Central Competitive Field except District 2 (central Pennsylvania). But non-union mines in Alabama, Kentucky, and West Virginia could erode Central Field markets.[41] Unable to win support for his own plan "to secure continuous employment and greater stability" for the industry, Hoover had hopes for the new coal commission. An expert inquiry led by "sound" men who could present facts to the nation and the Congress for action (like Wilson's earlier, smaller commission), Harding's proposed panel helped solve a difficult strike. Congress empowered the body "to investigate and ascertain fully the facts and condition" in the industry. Of the seven members Harding named on October 10, 1922, none were operators or miners.[42]

40. House Committee on Interstate and Foreign Commerce, *Coal*, pp. 27–39; *New York Times*, September 22, 27, October 12, December 10, 24, 30, 1922.

41. Brophy, "Elements of a Progressive Union Policy," pp. 188–90; Hammond, "Coal Commission Reports," p. 546; Fowler, pp. 101–103; Suffern, p. 157; Carnes, p. 101; Perlman and Taft, pp. 485–87; *New York Times* editorial, January 21, 1923; *United Mine Workers Journal*, August 15, 1922, pp. 3–7; September 15, 1922, p. 12; Anna Rochester, *Labor and Coal* (New York: International Publishers Co., 1931), pp. 201–202.

42. Parker, p. 91; Hammond, "Coal Commission Reports," p. 547; *New York Times*, August 19, 1922; February 11, 1923; Mary Van Kleeck, *Miners and Management: A Study of the Collective Agreement Between the United Mine Workers of America and the Rocky Mountain Fuel Company* (New York: Russell Sage Foundation, 1934), pp. 197–98; *United Mine Workers Journal*, October 1, 1922, p. 6; November 1, 1922, p. 6; September 1, 1922, pp. 6–7.

Led by John Hays Hammond, the eminent public figures on the commission failed to improve the industry. They solicited data from operators and miners and incorporated some excellent statistical tables in their final report. But preliminary reports, issued in mimeographed form, were scarcely mentioned by the press, much less Congress, which appropriated no money for printing the final report. Worse, the commissioners' attempt to reach a unanimous opinion in every area led to ambiguous conclusions.[43]

The commission's report criticized the "medieval methods" (yellow-dog contracts, spy systems, privately paid deputy sheriffs) used by some non-union employers in fighting the union but praised the vastly better living conditions in the union as opposed to the non-union camps. It called the labor-management difficulties in coal an "irrepressible conflict." But it refused to support the idea of unionization and instead exhorted the union to provide "service to the public under conditions fair alike to employer and employee. . . ."[44]

The report declared that much of the industry was a public utility, "clothed with a public interest," yet it placed "the largest responsibility for putting the coal industry in order" on the industry itself. The coal commission wanted the Interstate Commerce Commission to establish a coal division to discourage development of high-cost mines, cross-hauling, and long hauls and to allot railroad cars on the ability of the operator to sell, rather than merely produce, coal. Taxes would absorb excess profits, and licensing would monitor trade practices. The commissioners failed to decide on a unified approach to effect these excellent proposals and simply called for "better public understanding of the coal business." Coolidge, who had assumed the presidency by the time the commission's report was completed, called it "helpful and wise" but dropped the matter there. Beginning to show his rhetorical ability, John

43. Hoover, *Memoirs*, vol. 2, p. 70; Hammond, "Coal Commission Reports," pp. 548–50, 578; *New York Times*, January 16, 1923; Edward T. Devine, *Coal: Economic Problems of the Mining, Marketing and Consumption of Anthracite and Soft Coal in the United States* (Bloomington, Ill.: American Review Service Press, 1925), p. 408; United Mine Workers of America, *United Mine Workers of America and the United States Coal Commission* (Indianapolis: UMWA, 1923), pp. 1–2.

44. U.S., Coal Commission, *Report*, 5 parts, 68th Cong., 2d sess., 1925, part 3, p. 1333; Hammond, "Coal Commission Reports," pp. 574–77.

L. Lewis styled it a "maze of well-worn generalities, which could have been written by any well-informed mine superintendent within a sixty-day period." He suggested that it be "duly filed and the dust of the ages allowed to collect thereon." The product of one of the largest strikes in the industry, the commission turned out to be one of the greatest fiascos.[45]

The Coolidge Years: The Jacksonville Agreement and After

By 1924, miners, operators, and government officials sought a breathing spell from strikes, gyrations in prices, and government investigation. The economy briskly revived from the postwar recession. Perhaps Calvin Coolidge could bring simpler times. In part to avoid labor disturbance in an election year, Hoover, who had moderated his enthusiasm for improving the efficiency of the industry, pressed operators in the Central Competitive Field toward a status-quo wage contract in 1924. At a private dinner arranged by George G. Moore, a wealthy operator and utilities magnate, Hoover, John L. Lewis, a newsman, and two prominent men from the New York financial world worked out a plan to avert a strike. Under the constraints of the expanding southern non-union fields, Lewis agreed to seek no advances over the current $7.50 daily wage scale. Continuing the relatively high current wage scale would, Hoover and Lewis reasoned, benefit the mechanized Central Field mines and retard the growth of the inefficient non-union areas. In "a period of continuous operation under free competition and full movement of coal," Hoover mused, perhaps the industry might stabilize itself. In Jacksonville, Florida, the operators and representatives of the miners followed the Hoover-Lewis reasoning and negotiated a new three-year agreement reaffirming the $7.50-a-day basic wage, effective April 1, 1924.[46]

The Jacksonville Agreement marked the end of the post-

45. Parker, pp. 92–93; Hammond, "Coal Commission Reports," pp. 579–81; Soule, p. 178; United Mine Workers, *UMW and the USCC*, pp. 2–3; *New York Times*, September 26, 39, 1923; Hawley, "Hoover and the Coal Problem," p. 263.

46. Fowler, pp. 142–43; Walton Hamilton and Helen R. Wright, *The Case of Bituminous Coal* (New York: The Macmillan Company, 1925), pp. 237–38; Parker, p. 70; Devine, pp. 217–18; Hawley, "Hoover and the Coal Problem," p. 70; Murray, *Harding Era*, pp. 504–506, 513; Dubofsky and Van Tine, p. 107.

war efforts to bring order to the bituminous industry. Efforts to re-create the war programs died in the 1919 strike. The high prices and shortages of 1919 and 1921 brought consumer pressure for legislation but no action. The U.S. Coal Commission's recommendations went into limbo. Union operators left the National Coal Association and weakened it as a spokesman for national action. Insurgent attacks on Lewis began in Kansas, Illinois, and District 2.

Lewis's gamble at Jacksonville was based on his hope that the wage agreement would drive inefficient mines out of business and put some two hundred thousand surplus miners out of work. This scheme depended on the ability of the mechanized mines of the North to compete with the less efficient non-union southern mines, on the Interstate Commerce Commission's revision of freight rates to favor union-mined coal, and on government backing of the union position in the industry. None of Lewis's premises proved valid. Instead, government saw the non-union coal as a protection against Lewis's power to call a strike, and the union operators found they could not compete with southern coal. Central Field operators soon wanted to renegotiate the Jacksonville Agreement.

Heeding Lewis's policy of "no backward step," the union refused. Leasing arrangements designed to skirt the contract sprang up in Knox County, Indiana, and elsewhere. Operators in southeastern Ohio and western Pennsylvania opened up old mines at the 1917 pay scale. The giant Consolidation Coal Company cut daily wages to $6.00. When the Pittsburgh Coal Company, a leader in the Central Field which employed seventeen thousand men and had been a union operation for thirty-five years, shut down under changed management headed by the Mellon family and reopened as a non-union company in 1925, the Jacksonville Agreement crumbled. Pittsburgh Coal and Iron Company and Youghioghency and Ohio Coal Company followed Pittsburgh Coal Company's lead.[47]

47. NRA, "Economic Survey," pp. 1, 70, 74. Perlman and Taft, pp. 563–64; John L. Lewis, *The Miner's Fight for American Standards* (Indianapolis: UMWA, 1927), p. 37; Reed Moyer, *Competition in the Midwestern Coal Industry* (Cambridge, Mass.: Harvard University Press, 1964), p. 109; McDonald, pp. 167–74; Fowler, pp. 144–46; Dubofsky and Van Tine, pp. 132–36.

Lewis struck those in Pennsylvania who had defied the Jacksonville Agreement and triggered another bloody labor conflict. Pittsburgh Coal Company offered scabs liquor and prostitutes and also evicted striking miners, including one whose wife later died in the winter cold. The management of Pittsburgh Coal wrote local mine officials to keep company police in the background and to "clean up all unsightly conditions," but a senator found that "conditions which exist in the strike-torn regions of the Pittsburgh district are a blotch upon American civilization." He remarked: "Women and children [are] living in hovels which are more unsanitary than a modern swine pen. They are breeding places of sickness and crime." Thousands of women and children, reported the *New York Daily News* were "literally starving to death" under "a system of despotic tyranny reminiscent of Czar-ridden Siberia at its worst."[48]

Lewis claimed to have spent nearly eight million dollars on strikes to maintain the Jacksonville Agreement, but he could not resist the economic pressures on the industry. Between 1924 and 1927 the traditionally unionized fields of Pennsylvania, Ohio, Indiana, and Illinois lost forty-four million tons of yearly production. Non-union fields in Kentucky, West Virginia, and Virginia picked up fifty-seven million tons. Former union miners in northern camps went south to get work. Although the area covered by the Jacksonville Agreement gave men higher annual earnings, the union strength north of the Ohio River dwindled.[49]

Progressives tried to "save the union." Renewing his charge that Lewis must insist that union operators sign for all

48. U.S., Congress, Senate, Committee on Interstate Commerce, *Conditions in the Coal Fields of Pennsylvania, West Virginia, and Ohio, Pursuant to S. Res. 105,* 70th Cong., 1st sess., 1928, pp. 4, 11–17; Irving Bernstein, *The Lean Years: A History of the American Worker, 1920–1922* (Baltimore: Penguin Books, 1966; originally published by Houghton Mifflin, 1960), pp. 127–36; McDonald, pp. 177–78; Idaho Senator Frank Gooding quoted by Coleman, p. 132; Pittsburgh Coal Company letter quoted in Anna Rochester, *Rulers of America: A Study of Finance Capital* (New York: International Publishers Company, 1937), p. 70; *Fortune,* October, 1933, pp. 60–61.

49. McDonald, p. 174; NRA, "Economic Survey," vol. 1, pp. 64–76; Parker, p. 77; Waldo E. Fisher, *Collective Bargaining in the Bituminous Coal Industry: An Appraisal* (Philadelphia: University of Pennsylvania Press, 1948), p. 15; *New York Times,* September 22, 1933.

their mines as a condition for dealing with the UMW, Scot John Brophy launched his 1926 drive for the UMW presidency. Brophy's "Save the Union" platform proclaimed democracy in the union, organization of the non-union fields, a labor party, an educational campaign among miners for nationalization of the mines, and the six-hour day and five-day week. Assisted by the Communist Trade Union Educational League, but generally poorly funded, the Brophy drive sputtered. "Paper" districts loyal to Lewis—some of which had few duespayers—cast unanimous ballots for the regular Lewis slate. Lewis denounced Brophy's 1928 convention of his supporters in Pittsburgh as a dual union and expelled those who attended it from the UMW. Lewis's union was shrinking, but he kept control.[50]

In the Miami, Florida, wage talks in 1927, Lewis castigated the remnants of the union operators for their disorderly industry. They needed, he boomed, the union's stabilizing influence. But the coal men refused to reinstate the Jacksonville scale. Consumers had stockpiled eighty million tons of coal. Coolidge would not pressure the operators as Hoover had in 1924 for good reason: in 1927 the non-union fields could produce the coal needs of the nation. On the first day of the inevitable strike of 1927, a southern Ohio operator posted a notice stating that the "mines of West Virginia are today working full time filling your orders while you have no work." With new equipment those mines could

> take care of their own trade and also all the trade formerly held by the Ohio mines.
>
> Your officials say *You Men* instructed them to refuse a competitive scale. That means that *You Men* have forced your own mines to cease operations, and have *Given Away Our Trade and Have Lost Your Own Jobs to Favor West Virginia and the Nonunion fields.*
>
> Have the West Virginia operators given you men anything for selling out our trade and your jobs?

50. Transcript of oral history interview between John Brophy and John O. P. Hall, pp. 33–35, 49–50, 480; Brophy, "A Larger Program for the Miners' Union"; allegations of vote fraud with documentation, all in box for 1924–29, Brophy Papers; Perlman and Taft, pp. 564–67; Taft, pp. 392–99; Bernstein, p. 106.

The industry faced technological and other changes which were eroding its markets. Spurred by the development of the diesel engine, oil continued to invade the railroad fuel market during the 1920's just as the rate of growth of the railroad industry slacked off. Petroleum and natural gas replaced coal in many commercial markets. Between 1919 and 1929 the index of the output of manufacturers grew by over 50 percent, but coal's share per unit of output declined during the same time by 33 percent.[51]

Both economics and nature attacked the miners in 1927. The long strike fizzled. Lewis produced some district agreements at reduced wages in Illinois and Indiana, but Ohio and Pennsylvania went largely non-union. On May 30, 1927, torrential rains in Appalachia produced flash floods which devastated crops, homes, and mines. Miners in the Appalachian hills had built their homes in the hollows and farmed the hillsides. Overnight torrents of water washed the tilled hillsides into the valleys and carried all to destruction. At dawn, rescuers found nearly one hundred dead and railroad tracks ripped up by the force of the floods, which had carried tons of rock and huge slag piles into the mining camps. Denuded of their soil, the hills "lay bleak and dead in a state of sterility from which they have not recovered to this day."[52]

During the brief price rise brought by the 1927 strike, consumers pressed for relief and even nationalization "for the production and distribution of coal at reasonable prices." But when the country rode through the 1927 strike without the shortages of 1922, consumer spokesmen lost their audience. Hoover backed one bill to create a division of coal economics in the Commerce Department, but it died on the House floor. Now primarily non-union, the industry maintained its anti-

51. Mine notice quoted in Fowler, p. 150; *Coal Age* 30 (September 9, 1926): 441; Harold Barger and S. H. Schurr, *The Mining Industries, 1889–1939* (New York: National Bureau of Economic Research, 1944), pp. 77–80.

52. Hawley, "Hoover and the Coal Problem," p. 267; Fowler, p. 152; Parker, p. 72; Van Kleek, p. 51; McDonald, pp. 175–80; Harry M. Caudill, *Night Comes to the Cumberlands: A Biography of a Depressed Area* (Boston: Little, Brown and Co., 1962), pp. 144–53.

legislation stance. Lewis, however, planned ways to use government to reverse the laws of supply and demand.[53]

The Great Depression: The UMW Sponsors Stabilization Legislation

Sometime between the collapse of the Jacksonville Agreement and the spring of 1928, John L. Lewis broke the basic traditions of the American Federation of Labor and sought government help for the coal industry. A Republican to whom Coolidge had offered the secretaryship of labor and who would vote for Hoover in 1928, Lewis was in many ways more conservative than AFL founder Samuel Gompers. But Lewis jettisoned Gompers's principles of business unionism in the twenties just as he would the tenets of craft unionism later on. Aware that his union could not negotiate improved wage contracts, he sought a law to encourage operators to end wage and price cutting. Drawing on the ideas of the UMW's chief economist, W. Jett Lauck, Lewis wanted to replace marketplace competition which drove down wages and prices with a regulatory system which gave his union power to bargain for nationwide wage scales and gave the operators means to end price slashing. His plan would reward efficiency of operation. A new phase in the relationships among business, labor, and government in the bituminous industry had begun.[54]

Introduced by Senator James E. Watson, Indiana Republican, the UMW proposal was the obverse of earlier consumer-oriented legislation. All corporations producing coal for interstate commerce would be licensed and required to bargain collectively with their workers. In return they became exempt from the antitrust laws and could form selling associations or pools to halt price cutting. A five-man bituminous coal commission would investigate the industry, set

53. Hawley, "Hoover and the Coal Problem," pp. 265–66; Parker, pp. 95–97; *Coal Age* 29 (March 18, 1926): 405.
54. Alinsky, pp. 361–63; Sulzberger, p. 5; Carnes, p. 227; Dubofsky and Van Tine, pp. 137–38.

maximum prices if necessary, and oversee the creation of the selling associations. Using the wartime powers of the Fuel Administration, the ICC would limit railroad sidings to new mines. The bill protected miners' freedom of assembly and speech and their right to a checkweighman, and it banned company scrip. Selling asociations and collective bargaining, the union believed, would protect wages and restore the union.[55]

Operators and coal retailers united against the proposal. "The federal government will be underwriting profits of the soft-coal industry, and wages of employees engaged in that industry," prophesied the retailers, "and the public will foot the bill." There had been no year since 1902, complained H. E. Willard, secretary of United States Coal Co., that "our mines have not been closed in utter violation of the printed contract" with the UMW. The Hanna Coal properties in Ohio had been selling below cost for nearly three years, argued R. L. Ireland, and could not afford unionization and higher wages. Primarily a gesture to the UMW, the bill languished in committee to be reintroduced in 1929 with similar lack of success. Both 1928 party platforms backed the idea of aiding the coal industry, however, and the Watson bill sowed some of the seeds for the National Recovery Administration (NRA) of 1933.[56] Like the later NRA, the Watson bill aimed at getting the industry to increase wages while providing price protection, a solution that took no account of the long-term decline of the industry's share of the energy market.

Following the collapse of the 1929 stock market, the economy retrenched, and demand for coal declined. Although up slightly in 1929, bituminous use fell nearly 10 percent in 1930, another 18 percent in 1931, and another 20 percent in 1932. With the exception of the 1922 nationwide strike,

55. Theodore J. Kreps, *Business and Government under the National Recovery Administration* (New York: American Council Institute of Pacific Relations, 1936), p. 13; *Coal Age* 33 (June, 1928): 389; 34 (September, 1929): 564.

56. U.S., Congress, Senate, Committee on Interstate Commerce, *Bituminous Coal Commission: Hearing on S. 4490*, 70th Cong., 2d sess., 1929, pp. 1–4 and passim; Brophy, Oral History Memoir, pp. 753–54, Brophy Papers, Arthur M. Schlesinger, Jr., gen. ed., *History of American Presidential Elections, 1889–1968*, 4 vols. (New York: Chelsea House Publishers), vol. 3, pp. 2621, 2633; *Coal Age* 33 (June, 1928): 389–90; 34 (February, 1929): 123.

miners dug less coal in June, 1932, than at any other time since monthly records were begun in 1905. Not since 1904 had annual production fallen as low as it did in 1932. Considered a "sick" industry since the prewar days, bituminous coal was dying.[57]

All the destructive forces at work in coal since the mid-1920's accelerated. Oversupply of mines and miners led to price and wage cuts. Worth between $2.00 and $4.00 a ton in the postwar years, coal in 1926 had averaged $2.05 a ton, and it fell to $1.31 a ton in 1932. The $7.05 daily wage of the Jacksonville Agreement vanished. Some northern operators still recognized the union and paid between $3.00 and $6.00 a day, but over one-third of the industry paid a daily wage of less than $2.50. Some miners got $1.00 a day, and many $1.50. Coal's share of the energy market reached a low of 45 percent as users switched to petroleum and natural gas. More than thirty-five hundred mines closed, and approximately three hundred thousand miners lost their jobs. Those still employed at depression wages in 1932 worked only an average of 146 days.[58]

Hunger and misery stalked many mining camps. An epidemic of dysentery caused by malnutrition spread through the West Virginia mining areas, which were still suffering from the effects of the flood. "The fare of the workers and their dependents," Lewis wrote Hoover in 1932, "is actually below domestic animal standards." Miners' families ate "bulldog gravy" made of flour, water, and some grease. Without adequate clothing, children stayed out of school. In one school an inspector found many pupils who had never tasted milk. Of two hundred students, thirty had pellagra, and others had rickets and tuberculosis. Employed miners who mined tons of coal each day had to salvage a few lumps of kitchen fuel from the slag pile. "In the whole range of depression," Pennsylvania's Governor Gifford Pinchot em-

57. NRA, "Economic Survey," vol. 1, p. 35. *Bradstreet's* 40 (September 10, 1932): 1158; (August 13, 1932): 1031; Muriel Sheppard, *Cloud by Day: The Story of Coal and Coke and People* (Chapel Hill: University of North Carolina Press, 1947), p. 109.

58. Barger and Schurr, p. 163; Fowler, pp. 39–40; Parker, p. 73; NRA, "Economic Survey," vol. 2, p. 278; Morris, p. 22; Bernstein, p. 361.

phasized, "there is nothing worse than the condition of the soft coal miners." A group of social workers with European relief experience declared that the conditions in West Virginia "can be compared only with those in Poland and Russia following the war."[59]

Herbert Hoover, the Great Engineer who had labored hard to bring some kind of efficiency to bituminous coal as secretary of commerce, easily defeated Alfred E. Smith in the presidential election of 1928. Born at the end of the depression of the 1870's, Hoover had earned his humanitarian reputation. He had chaired the Commission for Relief in Belgium during World War I and distributed millions of tons of food to starving Europeans after the war. Personally heading a massive relief expedition into the Mississippi Valley during the flood of 1927, Hoover commandeered river steamers, spearheaded a Red Cross radio drive that netted fifteen million dollars, and organized nonprofit groups through the U.S. Chamber of Commerce to raise ten million dollars' worth of rehabilitation loans. A genuine American folk hero, a self-made millionaire from humble Quaker origins, World War I food administrator and for eight years secretary of commerce, Hoover entered the presidency with better preparation for the job than any other occupant since Washington.[60]

Although a member of both Harding's and Coolidge's cabinets, Hoover was much more progressive than either of those men. When the stock market crashed in 1929, he realized that the impact would be worldwide and so informed leading industrialists at White House conferences in the fall of 1929. He won pledges to maintain wages, prices, and

59. Tom Tippett, letter to the editor, *The Nation* 135 (July 20, 1933): 57; letter to the editor, *Christian Century* 46 (July 20, 1933): 913; Lewis quoted in Bernstein, *New Deal Collective Bargaining Policy* (Berkeley: University of California Press, 1950), p. 16; Morris, pp. 106–109; Bernstein, p. 363; Malcolm Ross, *Machine Age in the Hills* (New York: The Macmillan Co., 1933), p. 74; Pinchot quoted in the *New York Times*, January 31, 1932; Malcolm Ross, "Lifting the Coal Miner Out of the Muck," *New York Times Magazine*, October 1, 1933, pp. 4–5.

60. Hoover, *Memoirs*, vol. 2; Richard Hofstadter, *The American Political Tradition and the Men Who Made It* (New York: Alfred A. Knopf, 1948), chapter 11; Carl Degler, "The Ordeal of Herbert Hoover," *Yale Review* 52 (Summer, 1953): 563–83; Harris G. Warren, *Herbert Hoover and the Great Depression* (New York: Oxford University Press, 1959); James David Barber, *Presidential Character: Predicting Performance in the White House* (Englewood Cliffs, N.J.: Prentice Hall, Inc., 1973).

construction starts. Hoover tried to secure a moratorium on international debts. He expanded public works, unbalancing the federal budget, although he later signed the largest tax increase in peacetime to rectify this break with economic orthodoxy. Hoover also encouraged the Federal Reserve Board to expand credit by lowering the rediscount rate.[61]

But Hoover viewed public relief for unemployment with suspicion. Convinced that federal relief programs would both dry up private sources of money for charity and weaken the moral fiber of the recipients, Hoover tried to rely on private, state, and local relief programs. Headed by Council of National Defense veteran Walter S. Gifford, Hoover's Organization on Unemployment Relief attempted to stimulate private activity to help the unemployed. When the committee was created on August 19, 1931, from the remnants of the earlier Emergency Committee for Employment, Gifford declared that "between October 19 and November 25 America will feel the thrill of a great spiritual experience. In those few weeks millions of dollars will be raised in cities and towns throughout the land, and the fear of cold and hunger will be banished from the hearts of thousands!" The committee did increase giving to community chests, but it could not begin to assist the fourteen millions who were unemployed in 1932. Following a series of vetoes, Hoover signed the Wagner-Rainey Act in July, 1932, to allow financially exhausted states to borrow from a $300 million relief fund administered by the Reconstruction Finance Corporation.[62]

Hoover's reliance on private charity failed to relieve distress in the mining camps. The Organization on Unemployment Relief praised 1932 efforts by Consolidation Coal Company to stimulate employee gardening among its workers. By supplying seed packets, the company encouraged tilling of over four thousand gardens and sold Mason jars through the company stores so wives could put up the produce for the winter. In the hills of West Virginia, Quak-

61. See Degler, "Ordeal of Herbert Hoover"; Murray Rothbard, essay in Huthmacher and Sussman, pp. 43–53, and the rejoinders to Rothbard in the same volume; Warren, Chapters 9–12.

62. Leuchtenburg, "The New Deal and the Analogue of War," in *Change and Continuity in Twentieth Century America*, ed. John Braeman, et al. (New York: Harper and Row, 1964), pp. 97–98; Warren, chapter 13.

ers began handicraft industries and food programs. But the Friends found thousands of undernourished children and could feed only those who were seriously underweight. In one camp where there was no relief, sixteen miners told the county judge that they planned to break into the company commissary unless they received some aid. A visitor to Appalachia met a forty-three-year-old miner's wife, pregnant and mother of twenty, of whom only three had lived. She was hitchhiking to town to seek Red Cross aid.[63]

Resolute against federal relief, Hoover nonetheless believed in legislation to help the natural resource industries. "It certainly is not the purpose of our competitive system," Hoover remarked of the coal industry before the American Federation of Labor in October, 1930, "that it should produce a competition which destroys stability in an industry and reduces to poverty all those within it." He concluded, "If our regulatory laws be at fault they should be revised." In his 1930 state-of-the-union address he asked Congress to inquire into the effect of the antitrust laws on natural resource industries to see whether competition had become destructive and wasteful. Lewis and others "bombarded" him to generate operator-miner discussion on possible legislation, but Hoover refused to intervene personally. He referred the problem to the secretaries of labor and commerce to see if there was operator interest. In July, 1931, Commerce Secretary Robert Lamont met with some operators who claimed that they could "put their own house in order." Labor Secretary William N. Doak polled some 160 operators in 1931 on meeting jointly with union officials that September to devise legislation to alleviate the effects of the depression; he found only thirty-eight who would meet with the union. Without prior commitment from operators and miners, legislation seemed unlikely.[64]

63. *Business Week*, April 27, 1932, p. 25; Morris, pp. 105, 112–13; *Christian Century* 49 (August 31, 1932): 1054; Bernstein, p. 363; Ross, p. 111.
64. Hoover to Doak, July 21, 1930, August 13, 1930; Doak, memorandum for the president, "In re Proposed Bituminous Coal Conference," August 28, 1931; Doak to John L. Lewis, August 29, 1931, all in Coal, Presidential Subject File 104, Hoover Papers; *Congressional Record*, 71st Cong., 3rd sess., vol. 74, part 1, p. 35; *Commission*, p. 6; *Business Week*, July 15, 1931, p. 7; July 22, 1931, p. 9; September 9, 1931, p. 7; Parker, p. 101.

Governors of coal states stepped into the breach. Kentucky's Flem Sampson brought operators together to discuss allotment with inconclusive results. But a committee of five operators named by Pennsylvania's Pinchot and headed by Charles F. Hosford came up with a stabilization plan. Later to chair the New Deal's 1935 National Bituminous Coal Commission, Hosford set down four principles which started the operators thinking along the lines later developed under NRA. He advocated minimum wages, a five-day week, prohibition of sales below cost, and a code of fair trade practices. Some weeks earlier General Electric President Gerard Swope urged businessmen to use trade associations to fight the depression through better accounting, standardized trade practices, and shared information.[65]

Using fair trade codes to stabilize the industry had appeal. Encouraged by the experiences of their WIB-oriented war service committees and the FTC conferences on fair trade practices of 1919, many industries had organized trade associations to develop rules of fair trade. Once the FTC Division of Fair Trade Practices was established in 1926, more than 150 industries, including coal operators in Illinois, Indiana, and eastern Ohio, adopted codes.[66]

Hosford's proposal, however, came in the middle of a campaign against restrictive codes by Hoover's antitrust chief, John Lord O'Brian. Under pressure from O'Brian, FTC Chairman William Humphrey in 1930 began to force revisions of the existing fair trade codes believed to violate the antitrust statutes. Industrialists changed more than sixty codes, including one in soft coal. Hoover refused to repudiate the efforts of the FTC and the Justice Department. Under the circumstances, the Justice Department might attack Hosford's proposal as a conspiracy in restraint of trade.[67]

65. *Business Week,* October 21, 1931, pp. 13–14; *To Create a Commission,* pp. 1186–87; *Coal Age* 36 (September, 1931): 468–72.

66. Soule, p. 141; George B. Galloway, *Industrial Planning under the Codes* (New York: Harper and Brothers, 1935), pp. 21–23; Charles F. Roos, *NRA Economic Planning* (Bloomington, Ill.: Principia Press, 1937), pp. 9, 14; Mitchell, pp. 230–32. For a full-scale treatment of trade-association activity in the 1920's, see Himmelberg, *Origins of NRA.*

67. *Business Week,* February 19, 1930, pp. 22–24; April 16, 1930, p. 25; Hoover, *Memoirs,* vol. 2, pp. 168–74; Ellis W. Hawley, "Herbert Hoover and American Corporatism, 1929–1933" (manuscript in my possession), pp. 5–7.

Another proposal—also dangerously close to violating the antitrust laws—caught on in 1931. Led by astute leaders like James D. Francis of Island Creek Coal Co., Virginia, West Virginia, Kentucky, and Tennessee coal producers formed Appalachian Coals, Inc., a regional sales agency. Controlled by the members, who received common stock based on their tonnage output, Appalachian sold coal on a commission basis and assigned orders to particular mines by percentages. Although Appalachian Coals eventually marketed 73 percent of the southern Appalachian coal, it did not dominate any sales market. The Justice Department took Appalachian Coals to court to determine whether the Supreme Court demanded a strict application of the antitrust laws even for depressed industries.[68]

While operator groups awaited a verdict on regional marketing agencies, John L. Lewis renewed his drive for a coal stabilization bill. On January 12, 1932, Pennsylvania Senator James J. Davis introduced a union-supported update of the 1928 Watson bill. A coal commission of five would license operators producing coal for interstate shipment and would encourage them to form marketing agencies to set minimum prices. Members would have to bargain collectively with their employees. The commission would pass on the prices set by the operators in their marketing pools. Clyde Kelly from Pittsburgh introduced a companion bill into the House. Anticipating NRA chief Hugh S. Johnson's theme by over a year, Kelly asserted that he and Davis "desire to give them [the operators] self-government to the very last degree but with an umpire in the public interest."[69]

Angered by the opposition testimony of large coal consumers and by the arguments from an NCA witness at congressional hearings that the bill was "the most radical depar-

68. Parker, chapter 10; *To Create a Commission*, pp. 513, 774, 1185, 1190–91, 1311; *Coal Age* 37 (January, 1932): 34; *Nation's Business* 21 (May, 1933): 24; *Bradstreet's* 40 (January 23, 1932): 98; *Business Week*, December 16, 1931, p. 18; Himmelberg, *Origins of NRA*, pp. 151–54.

69. Testimony of Kelly, William Green, Henry Warrum, UMW legal draftsman, and John L. Lewis in *To Create a Commission*, pp. 12–27, 61–80, 707–708; Lorwin, p. 456; Harris, p. 98; *Coal Age* 37 (February, 1932): 83; *Business Week*, August 19, 1931, p. 15; James Mark, president, UMW District 2, to Philip Murray, April 16, 1932, Box for 1930–33, Brophy Papers.

ture from our historic policies that has ever been offered,"
John L. Lewis vividly portrayed the struggle of sixteenth-
century English miners to organize for "shorter hours and
more wages and improved conditions." With their petition
denied, the workers struck, and "the masters brought strangers to take their places." When the miners attacked the re-
placements, "the King sent troops to protect the strangers
and to drive the mine workers away from their jobs and their
homes." Only a few days before, Lewis noted angrily, this
process had been repeated in Ohio. Convinced that his example demonstrated the "lack of organization in the coal
industry through the centuries," Lewis berated the operators as

> ... the most inefficient group engaged in American business
> enterprises. They have practically no form of organization.
> They have no code of ethics. They are simply engaged in a
> struggle to continue their existence and remain in business.
> Why, the larger interests of the country are preying upon the
> coal industry, buying its product at less than the cost of pro-
> duction, and compelling the operator to sell the blood and
> sinew and the bone of the hundreds of thousands of men who
> are engaged in the industry, and to sell the future of their
> children with their blood and their bone.[70]

Pennsylvanian Charles Hosford spoke in favor of the principles of the bill. *Bradstreet's* endorsed it. But non-union
operators, now dominant within the councils of the National
Coal Association, vigorously opposed the "compulsory unionization" measure. Operators scoffed when Senator Davis
said that collective bargaining protection did not necessarily
mean recognition of the UMW. Most operators agreed with a
man from Indiana, Pennsylvania, that they had "constitutional rights to hire a man and pay him what he agrees to
sell his services to me for." Under the practiced hand of
NCA coordinators, witnesses against the proposal made a
strong case in the Senate hearings. The Davis-Kelly bill,
wrote *Coal Age,* was "unionization by fiat" and an "undesir-

70. Lewis testimony in *To Create a Commission,* pp. 414–15, 1308–1309.

able extension of government activity in the field of private enterprise."[71]

In an extensive brief, Judge H. D. Rummel of West Virginia presented the NCA's constitutional views. First, congressional power over interstate commerce did not extend to mining, an intrastate activity. Second, no part of the industry was "affected with a public interest," and thus it could not be regulated as a public utility. Third, depriving an operator of his property rights by requiring a license to operate exceeded the power of Congress. Fourth, Congress could not delegate regulatory power to a commission without providing strict standards of operation. Finally, the collective bargaining sections violated the right of contract. John Lord O'Brian, now Hoover's attorney general, and Commerce Secretary Robert P. Lamont refused to issue a ruling on constitutionality, but Secretary of Labor Doak supported the bill. The majority of the Senate Mines and Mining Subcommittee declared the measure unconstitutional, ending consideration for 1932.[72]

Following the presidential election, Maryland Senator David J. Lewis, former coal miner and government director of the telephone and telegraph industries during the war, and Arizona Congressman Carl Hayden sponsored a bill modeled on the British Coal Mines Act of 1930 and the Fuel Administration's experience. Elective district boards of operators would be empowered to fix minimum prices, determine mine quotas, and oversee the sale of coal through sales agencies like Appalachian Coals, Inc. A coal operators' national council, elected by operators from their districts, would allocate production by district and review the minimum prices. The bill failed to face coal's decline in the energy market, included the labor protections of the Davis-

71. In ibid., see testimony of B. M. Clark, p. 455; Charles Hosford, pp. 113–25; E. L. Greever, pp. 406–16, 1184, and passim; Howard Eavenson, p. 511; and James J. Davis, pp. 29ff., 455. *Coal Age* 37 (March, 1932): 91; (April, 1932): 168; (May, 1932): 200; (June, 1932): 221–22; *Business Week*, January 17, 1932, p. 11; *Bradstreet's* 40 (February 20, 1932): 240.

72. Testimony of Judge H. D. Rummel and Senator Carl Hayden in *To Create a Commission*, pp. 75, 109–10, 250–342; Parker, p. 102; Irving Bernstein, *Turbulent Years: A History of the American Worker, 1933–1941* (Boston: Houghton Mifflin Co., 1970), p. 22; Doak, "Memorandum for the President," February 17, 1922, Presidential Subject File 104, Coal, Hoover Papers; *Coal Age* 37 (May, 1932): 211–12.

Kelly bill, and was being reconsidered in the spring of 1933 when the National Industrial Recovery Act preempted it.[73]

Despite his earlier enthusiasm for coal marketing cooperatives and his interest in modifying the antitrust laws in natural resouce industries, Hoover refused to back either the Davis-Kelly or the Lewis-Hayden proposals because they were too coercive. He followed a similar course in opposing rigid production restrictions for petroleum. He may have hoped that a court victory for Appalachian Coals, Inc., might give the industry the means to stabilize itself. He did fear that trade associations' fixed prices would destroy competition. Of the Swope plan, which encouraged similar trade association activity, Hoover wrote: "It means the repeal of the entire Sherman and Clayton Acts, and all other restrictions on combinations and monopoly. In fact, if such a thing were ever done, it means the decay of American industry from the day this scheme is born, because one cannot stabilize prices without protecting obsolete plants and inferior managements." When the National Industrial Recovery Act adopted segments of the Davis-Kelly bill and encouraged direct price fixing through trade associations, Hoover cried, "fascism."[74]

Conclusions

Many planted the seeds for the New Deal programs for coal. During the period between the war and Roosevelt's

73. David J. Lewis testimony in *To Create a Commission*, pp. 210–29; Parker, pp. 103–104; *Coal Age* 37 (March, 1932): 123; (July 1932): 282; *New York Times*, May 3, 1933; *Business Week*, January 18, 1933, p. 7; Gerald D. Nash, *United States Oil Policy, 1890–1964* (Pittsburgh: University of Pittsburgh Press, 1968), pp. 98–111; Philip Murray to John T. Jones, UMW representative, May 1, 1933, Box for 1930–33, Brophy Papers; W. Jett Lauck, "Coal Labor Legislation: A Case," *Annals* 184 (March, 1936): 130–32; *United Mine Workers Journal*, August 8, 1918, p. 16.

74. Herbert Hoover, State of the Union Address, *Congressional Record*, 72d Cong., 2d sess., vol. 76, part 1, p. 53; Hoover office memorandum on the Swope Plan, September 17, 1931, quoted in William Starr Myers and Walter H. Newton, *The Hoover Administration: A Documented Narrative* (New York: Charles Scribner's Sons, 1936), p. 119; Hawley, "Herbert Hoover and American Corporatism, 1929–1933," p. 21; Rothbard, essay in Huthmacher and Sussman, p. 53; *Coal Age* 37 (January, 1932): 34. Hoover's own plan for the coal industry was based on competition among cooperatives and elimination of excess mines. Hoover, "Plan to Secure Continuous Employment." For a lucid explanation of Hoover's ideas regarding the Swope Plan see Joan Hoff Wilson, *Herbert Hoover: Forgotten Progressive* (New York: Little, Brown and Co., 1975), pp. 151–56.

election, Garfield's plan, Kenyon's code, Hoover's plan, Appalachian Coals, Inc., and the union's legislative measures together made up the essentials of the NRA code program and much that became the Guffey acts of 1935 and 1937. Indeed, proposals for the coal industry were significant precursors for the NRA. But only the UMW, now the sole spokesman for the entire industry, backed legislative action. None of the legislative remedies got beyond hearings. Ten years elapsed between Kenyon's code proposal and the NRA codes. Most coal operators clung to the laissez-faire tradition. It took a war for Wilson and Garfield to stabilize the bituminous industry. It would take a depression, a new administration, and a revived union to repeat the process.

Thus, the bituminous industry did not capture or control government in the era of "Republican ascendancy" except in the negative sense that most operators resisted government intrusion in their affairs. During the early part of the decade, many governmental officials thought that the coal industry was a price-gouging monopoly and renewed the struggle between "the people" and the "interests." But this illusion of monopoly power was soon shattered as the wartime economic overexpansion forced prices down and eroded wages. During the late 1920's the industry's North-South split widened. Some of the more organization-minded Central Field operators warmed to the proposals of price stabilization devised by the UMW. But the operators' antigovernment, antiunion sentiment—concentrated in the South but prevalent throughout the industry—destroyed any hope of business-government cooperation between the First World War and the New Deal.

5

Drafting a Charter for Industrial Self-Government

ASSUMING OFFICE in a crisis more serious than that of 1917, Franklin Roosevelt pinned recovery hopes on a plan for industrial self-government modeled in part on the experience of the Wilsonian war years. The National Industrial Recovery Act (NIRA), called by Thomas Cochran "the most important attempt to fuse governmental and private industrial controls in American history," drew on many sources. But in both ideology and personnel it had roots in the war. The Fuel Administration experience and the coal proposals of the 1920's and early 1930's influenced the drafting of the NIRA, yet the fragmented coal industry balked at agreeing to a unified National Recovery Administration (NRA) code of fair competition.[1] Deep-seated hostility of the operators toward government control, regional rivalries, antiunionism, and initial government weakness made it difficult for the New Dealers to get the industry to join the NRA.

1. Thomas C. Cochran, *Business in American Life: A History* (New York: McGraw-Hill Book Co., 1972), p. 317; William E. Leuchtenburg, "The New Deal and the Analogue of War," in *Change and Continuity in Twentieth Century America*, ed. John Braeman et al. (New York: Harper and Row, 1964), p. 117. U.S., Congress, Senate, Committee on Mines and Mining, *To Create a Bituminous Coal Commission: Hearings before a Subcommittee*, 72d Cong., 1st sess., 1932, p. 824 (hereafter cited as *To Create a Commission*); Ellis W. Hawley, *The New Deal and the Problem of Monopoly* (Princeton: Princeton University Press, 1966), p. 28; U.S., Special Industrial Recovery Board (hereafter cited as SIRB), "Proceedings" (mimeographed), September 6, 1933, p. 2. I will use the initials NRA to refer to the agency created to administer the act, which I will abbreviate NIRA.

In the final analysis the United Mine Workers of America under John L. Lewis, who was himself under pressure from rebellious wildcat strikers in Pennsylvania, won a national wage agreement which helped the NRA obtain operator acceptance of an NRA coal code. NRA chief Hugh S. Johnson, whom Lewis befriended during the summer of 1933, and President Franklin Roosevelt repeatedly intervened in the code-drafting process to urge non-union operators to sign a nationwide code. In short, speaking for "the public," Roosevelt, Lewis, and some Central Field operators corralled the divided coal industry into signing an NRA code of fair competition.

Industrial Self-Government Revived

When set against the spectacle of mass unemployment, hunger marches, near riots in the farm belt, and the Hoover administration's eviction of World War I veterans from the nation's capital, the election campaign of 1932 presented a curious spectacle. Buttressed by a Brain Trust of men of both conservative and liberal persuasion, Roosevelt oscillated on many crucial issues, including the tariff and fiscal policy. He attacked Hoover's Farm Board for "advising farmers to allow 20 percent of their wheat lands to lie idle, to plow up every third row of cotton and to shoot every tenth dairy cow." In Detroit, the New York governor advocated removing the causes of poverty, "but he refused to spell out the methods because it was Sunday and he would not talk politics." Nominated by a reluctant GOP, Hoover ran a defensive campaign until October, when he vituperatively attacked Roosevelt. Hoover punched holes in Roosevelt's pronouncements but failed to elicit support for his own obviously inadequate programs. Although buoyed by a brief economic upturn during the summer, the president, exhausted by the burden of office, could hardly maintain his feet while speaking.[2]

2. Roosevelt quoted by Frank Freidel, *Franklin D. Roosevelt: Launching the New Deal* (Boston: Little, Brown, & Company, 1973), pp. 139–45, 318; Harris Gaylord Warren, *Herbert Hoover and the Great Depression* (New York: Oxford University Press, 1959), chapter 17; Leuchtenburg, *Franklin D. Roosevelt and the New Deal, 1932–1940* (New York: Harper and Row, 1963), chapter 1.

The leadership of the United Mine Workers divided over the candidates. John L. Lewis publicly stayed with Hoover, although he agreed with a Roosevelt agent to quietly and privately aid the challenger. UMW Vice-President Philip Murray and others among the top officials openly supported Roosevelt. In a Hyde Park conference, Roosevelt listened to Murray's arguments for government help for the industry. In several minor addresses in Indianapolis and Terre Haute, Roosevelt promised to call a conference of operators and miners to see if they could unite to stabilize the industry. If this failed, he promised to seek special legislation. Throughout the long campaign, however, Roosevelt made no reference to governmental protection for labor's right to organize and bargain collectively.[3]

Despite his vacillation, Franklin Roosevelt appeared sympathetic to labor's needs. He came out for shortening the work week, government employment exchanges, unemployment insurance, federal relief programs, public works, and a vague kind of economic planning to prevent depressions. Yet he also pledged to reduce government spending by 25 percent and attacked Hoover's piling "bureau on bureau, commission on commission." One miner noted that Roosevelt's "hcart seems to be in the right place as far as labor is concerned. . . . I'm going to vote for him. Anything to get rid of Hoover." Coal diggers in the Cumberland Plateau placed a sign atop a freshly covered privy:

> Here Lies Hoover,
> Damn his soul,
> Buried in a honey-hole
> Let him lay here till the end
> Poor man's enemy
> Rich man's friend.[4]

3. Melvin Dubofsky and Warren Van Tine, *John L. Lewis* (New York: Quadrangle/The New York Times Book Co., 1977), p. 177; Irving Bernstein, *New Deal Collective Bargaining Policy* (Berkeley: University of California Press, 1950), p. 27; testimony of Philip Murray in U.S., Congress, Senate, Committee on Interstate Commerce, *Stabilization of the Bituminous Coal Mining Industry: Hearings before a Subcommittee of the Committee on Interstate Commerce*, 74th Cong., 1st sess., 1935, p. 157; James Wechsler, *Labor Baron: A Portrait of John L. Lewis* (New York: William Morrow and Co., 1944), p. 133; John Brophy, Oral History Memoir (typescript), chapter 19, in Brophy Papers.

4. McAlister Coleman, *Men and Coal* (New York: Farrar and Rinehart, 1943), p. 145; Brophy, p. 502; Warren, p. 255; young miner quoted in Coleman, p. 145;

Roosevelt gave the fleeting impression he might support a general industrial stabilization measure along the lines of the UMW's coal bills of 1928, 1929, and 1932. He told businessmen assembled in San Francisco's Commonwealth Club that government must enter the economy and that "the responsible heads of finance and industry instead of acting each for himself, must work together to achieve the common end." He went on, "Whenever the lone wolf . . . whose hand is against every man's, declines to join in achieving an end recognized as being for the public welfare, and threatens to drag the industry back to a state of anarchy, the Government may properly be asked to apply restraint." Elsewhere Roosevelt recalled the war mobilization:

> Compare this panic-stricken policy of delay and improvisation with that devised to meet the emergency of war fifteen years ago.
>
> We met specific situations with considered, relevant measures of constructive value. There were the War Industries Board, the Food and Fuel Administration, the War Trade Board, the Shipping Board and many others.[5]

Roosevelt tacked left and right, but he did call for energetic, positive government. Contrasted with Hoover's statements and the vacuous Democratic platform, Roosevelt at times seemed singularly original and daring. In a prenomination address to students at Atlanta's Oglethorpe University, he decried the social waste of educating young people for professions overfilled because of lack of planning. A laissez-faire attitude toward "our great economic machine," said Roosevelt, "requires not only greater stoicism, but greater faith in immutable economic laws and less faith in the ability of man to control what he has created than I, for

Irving Bernstein, *Turbulent Years: A History of the American Worker, 1933–1941* (Boston: Houghton Mifflin Co., 1970), p. 3; poem quoted in Henry M. Caudill, *Night Comes to the Cumberlands: A Biography of a Depressed Area* (Boston: Little, Brown, and Co., 1962), pp. 183–84.

5. Franklin D. Roosevelt, *The Public Papers and Addresses of Franklin D. Roosevelt*, 3 vols., ed. Samuel I. Rosenman (New York: Random House, 1938–1950), vol. 1, pp. 624–25, 631–32, 742–56. Too much can be made of Roosevelt's Commonwealth Club speech, which he neither wrote nor studied before delivery. Carl Degler, "The Ordeal of Herbert Hoover," *Yale Review* 52 (Summer, 1963): 583.

one, have." He extolled "planning the creation and distribution of those products which our vast economic machine is capable of yielding. . . . The country needs, and unless I mistake its temper, the country demands bold, persistent experimentation."[6]

Following an impressive electoral victory and a tense interregnum marked by failure to cooperate by both president and president-elect, Roosevelt assumed power in one of the dramatic moments in America's history. On March 4 the nation seemed to have come to a full stop. In nearly every state the banks were either closed or operating under state restrictions. The New York Stock Exchange shut down; so did the Chicago Board of Trade. As Walter Lippmann wrote, "At the beginning of March the country was in such a state of confused desperation it would have followed almost any leader anywhere he chose to go."

On that overcast March day, thousands of citizens stood before the magnificent Capitol rotunda, bundled against the winds which whipped the ivy and flags on the inaugural stand. Hatless and coatless, Roosevelt excoriated the "rulers of the exchange of mankind's goods [who] have failed through their own stubbornness. . . . They have no vision, and when there is no vision, the people perish."[7]

Roosevelt and his New Dealers swept into a Washington which seemed to one observer like "a beleaguered capital in war time." During and after the famous Hundred Days, former war administrators arrived to head agencies modeled on the Wilsonian war mobilization. The Civilian Conservation Corps seemed more like a military than a civilian organization. The Agricultural Adjustment Act both revived war techniques and was run by a war administrator, George Peek, after former War Industries Board chief Bernard Baruch refused the job. The TVA, the banking holiday, the

6. Arthur M. Schlesinger, Jr., gen. ed., *History of American Presidential Elections, 1789–1968,* 4 vols. (New York: Chelsea House Publishers, 1971), vol. 3, p. 2742; Roosevelt, vol. 1, pp. 639–46.
7. Leuchtenburg, *Franklin D. Roosevelt and the New Deal,* pp. 38–40; James MacGregor Burns, *Roosevelt: The Lion and the Fox* (New York: Harcourt, Brace and World, 1956), pp. 163–66; Walter Lippmann, in *Review of Reviews,* May, 1933; Roosevelt, vol. 2, pp. 14–15.

economy measures all "shared in the war legacy." Isodor Lubin, a New Dealer who had served in the war mobilization, wrote, "The hotels are filled, and the restaurants remind me very much of war times. One cannot go into the Cosmos Club without meeting half a dozen persons whom he knew during the war."[8]

A group of coal men from Illinois and Indiana visited Secretary of the Interior Harold Ickes to urge him to support a bill to restore some kind of competitive equality between their districts and more distant fields with lower wage scales which were competing in Chicago markets. Endorsed later by the UMW, their proposal would have re-created a modified Fuel Administration. "They wanted the Government to set a minimum wage, fix minimum and maximum prices for the coal industry," Ickes wrote in his diary, "and if necessary limit the production of coal and prorate such production among the different states and coal fields." Some western Pennsylvania operators and Ohioan Frank E. Taplin proposed that a coal czar take over the industry. Another group of operators tried to sell their mines to the government, "at any price fixed by the government. Anything so we can get out of it."[9]

Operators outside the Central Competitive Field favored legal protections for sales agencies like Appalachian Coals, Inc., and feared government intervention. Shortly after the election, the Governmental Relations Committee of the National Coal Association pressed for a bill to exempt the natural resource industries from the antitrust laws and to encourage cooperative marketing schemes like those tried in agriculture after the First World War. Headed by Pennsylvanian Charles O'Neill, president of Peale, Peacock, and Kerr, the Governmental Relations Committee staunchly opposed the

8. Lubin and observer quoted by Leuchtenburg, "New Deal and the Analogue," pp. 123, 125; see also pp. 104–10.
9. Sydney Hale, editor of *Coal Age*, to F. E. Berquist, November 14, 1935, NRA, Consolidated Files on Industries Governed by Approved Codes (henceforth cited as CF), Bituminous Coal (henceforth cited as BC), Division I, Reports, RG 9, NA; *Coal Age* 38 (May, 1933): 168; Harold Ickes, *The Secret Diary of Harold Ickes*, vol. 1, *The First Thousand Days, 1933–1936* (New York: Simon & Shuster, 1953), pp. 24, 30–34; Frances Perkins, *The Roosevelt I Knew* (New York: Simon and Shuster, 1946), p. 230.

UMW-sponsored legislation and the Norris LaGuardia Act. The March, 1933, Supreme Court decision upholding the constitutionality of Appalachian Coals, Inc., convinced many in the Appalachian region that operator-controlled marketing agencies by themselves could stabilize the industry. The industry thus divided over a recovery plan.[10]

Two individuals on the NCA's Governmental Relations Committee, however, emerged as temporary leaders for the industry. A miner at eleven, Charles O'Neill became president of UMW District 2 and then owner of a small mine. He later led other central Pennsylvanians to form the Central Pennsylvania Coal Producers Association, of which he was general secretary for over a decade. A Catholic, a Democrat, and a blunt, straightforward man with definite sway among his fellows, O'Neill served on the NCA committee with the president of West Virginia's mammoth Island Creek Coal Company, James D. Francis. A founder of Appalachian Coals, Inc., and a recognized industry leader, Francis had angered John L. Lewis by blocking unionization at Island Creek for years, yet he was much less parochial than many West Virginia operators.[11]

In late March, 1933, with Ickes and Secretary of Labor Francis Perkins, whose appointment pleased women more than it did organized labor, Roosevelt heard UMW President John L. Lewis, Congressman David J. Lewis of Maryland, and Senator Carl Hayden of Arizona argue for creation of a national council of coal operators to prorate coal by districts. In exchange for granting collective bargaining, operators would set minimum prices and mine quotas through locally elected district boards. In the Guffey-Snyder Act of 1935, Roosevelt would support this kind of plan. In 1933 he encouraged the two secretaries to continue conferences with the UMW. But for the moment he withheld support.[12]

10. *Coal Age* 37 (February, 1932): 84; (December, 1932): 423, 450; 38 (January, 1938): 28; Ickes, p. 21.

11. Raymond Salvati, *"Island Creek": A Career Company Dedicated to Coal!* (New York: Newcomen Society in North America, 1957), p. 11; John L. Lewis's testimony in *To Create a Commission*, p. 1325.

12. Friedel, pp. 155–59; Ickes, p. 10; *Coal Age* 38 (April, 1933): 133; Raymond Moley memorandum to Marvin McIntyre, March 21, 1933, Official File 175, Box 1, Roosevelt Papers.

Despite the call for action in his inaugural address, Roosevelt had not originally intended to press for a major industrial recovery bill during the special session of Congress he summoned to deal with the banking crisis. But proponents of various schemes bombarded him with suggestions, and once he scored early successes with a receptive legislature, he sought a bill parallel to the Agricultural Adjustment Act to invigorate the national economy. Then, when Senator Hugo Black's proposal to spread work by shortening the work week to thirty hours passed the Senate on April 6, Roosevelt offered an omnibus national industrial recovery bill as an alternative.[13]

The recovery bill pulled together a multitude of proposals, including ideas from the specific postwar legislative measures offered to stabilize coal. Senator Robert Wagner of New York and Wisconsin's Robert La Follette pressed for expansion of public works. Old Bull Moose supporters like Chicago labor lawyer Donald Richberg and Columbia economist Rexford Tugwell sought government direction of a more concentrated economy. "Emergency governmental control," said Richberg to one senator, "is now as essential to the national welfare as it would be in a time of war...." Supported in part and somewhat reluctantly by fellow Brain Trusters Raymond Moley and Adolph Berle, and enthusiastically by Jerome Frank in the Department of Agriculture, Tugwell sought a centrally planned economy. But Roosevelt never seriously contemplated such a major economic innovation, nor would he sanction the massive federal spending required to revive the economy.[14]

13. In *The Origins of the National Recovery Administration* (New York: Fordham University Press, 1976), Robert F. Himmelberg offers the most thorough treatment of the origins of NIRA. He emphasizes the impact of those seeking revision of the antitrust laws. Freidel, chapter 24; Burns, p. 168; J. Joseph Huthmacher, *Senator Robert F. Wagner and the Rise of Urban Liberalism* (New York: Atheneum, 1968), p. 144; Bernstein, *Turbulent Years*, p. 24.

14. Richberg quoted by Arthur M. Schlesinger, Jr., *The Age of Roosevelt*, vol. 2, *The Coming of the New Deal* (Boston: Houghton Mifflin, 1958), pp. 92–94; Leuchtenburg, *Franklin D. Roosevelt and the New Deal*, pp. 56–57; Charles F. Roos, *NRA Economic Planning* (Bloomington, Ill.: Principia Press, 1937), pp. 4–5, 35; *To Create a Commission*, p. 756; John L. Lewis, "Labor and the NRA," *Annals of the American Academy* 172 (March, 1934): 58; Bernstein, *Turbulent Years*, p. 23; Himmelberg, pp. 182–83.

The more visionary proponents of government intervention and control were matched by a host of skillful men urging revision of the antitrust laws so trade associations could rebuild the economy through "business planning." Walker D. Hines, former director general of the Railroad Administration, in 1933 head of the Cotton Textile Institute, impressed Roosevelt with the need for cooperation among industrialists in the highly competitive cotton textile industry. WIB veterans Gerard Swope of General Electric and General Hugh S. Johnson, the National Association of Manufacturers, and Henry I. Harriman of the Chamber of Commerce all endorsed the trade association concept. Franklin D. Roosevelt had been a president of the American Construction Council, the trade group for the construction industry, but he initially had turned a deaf ear to revisions of the antitrust laws as a recovery device. Once the trade association supporters mobilized themselves and their arguments, however, they would take advantage of circumstances.[15]

Ray Moley, appointed by Roosevelt to act as a kind of clearing house for recovery measures, one day bumped into General Hugh S. Johnson at the Carlton Hotel and urged him to work up a bill. "Nobody can do it better than you," said Moley. "You're familiar with the only comparable thing that's ever been done—the work of the War Industries Board." Assisted by Donald Richberg on the labor provisions, Hugh Johnson put together a draft that empowered the president to approve trade association agreements, revoked the antitrust laws, and used the federal licensing power to force compliance. "I never went back to New York from that day to the end of my service," the impetuous former cavalry officer later wrote, "except to get my clothes. . . ." Indeed, Johnson's foibles, bombastic statements, and feverish activity would come to epitomize the early New Deal.[16]

15. Hawley, pp. 42–43; Roos, pp. 5–7; Bernstein, *Turbulent Years*, pp. 19–20; Louis Galambos, *Competition and Cooperation: The Emergence of a National Trade Association* (Baltimore: Johns Hopkins University Press, 1966), pp. 193–95; Gerard Swope, *The Swope Plan: Details, Criticisms, Analysis*, ed. J. George Frederick (New York: The Business Bourse, 1931); *New York Times*, May 4, 6, 1933; Himmelberg, chapter 11.

16. Hugh S. Johnson, *The Blue Eagle from Egg to Earth* (New York: Doubleday, Doran and Co., 1934), pp. 193, 196.

Simultaneously, men around Senator Wagner hammered out a bill to expand public works and protect labor. In the Wagner group, labor, particularly representatives from the United Mine Workers, played a significant role. The roots of what later became Section 7a of the NIRA, defining collective bargaining rights for labor, extended deep into the legislative history of the coal industry. Iowa Senator William Kenyon's industrial code, Watson's coal stabilization bills of 1928 and 1929, and the Davis-Kelly bill of 1932 all had language parallel to that of Section 7a. John L. Lewis bragged that at Bernard Baruch's New York home, when General Johnson, Baruch, and assorted industrialists discussed recovery plans, "I was the only representative of organized labor present. I insisted on the inclusion of what later came to be known as Section 7a into this piece of legislation. I fought for it and I got it."[17]

Lewis exaggerated the significance of this single meeting, but the UMW did help win inclusion of Section 7a in the NIRA. Meyer Jacobstein, a Brookings Institution economist and former congressional author of a 1926 coal bill; Dr. W. Jett Lauck, a draftsman of the coal measures of the late 1920's; and Pennsylvania Congressman Clyde Kelly, the sponsor of the House coal stabilization bill of 1923, all worked on Wagner's drafting committee. Lauck recalled that the future Section 7a "was taken verbatim from the Davis-Kelly Bill. I know this because Congressman Clyde Kelly and I did this in drafting the Recovery Act."[18]

In the Department of Commerce another group headed by Assistant Secretary John Dickinson and influenced by H. I. Harriman and other business leaders merged with the Wagner group. Although agreement between the Wagner-Dickinson draftsmen and the Johnson-Moley faction would

17. See chapter 4. Bernstein, *Collective Bargaining*, p. 24; Saul Alinsky quoted Lewis in his *John L. Lewis: An Unauthorized Biography* (New York: G. P. Putnam's Sons, 1949), pp. 67–68; William Jett Lauck, UMW economist, to John T. Flynn, April 27, 1936, Correspondence—NRA, Lauck Papers.

18. Huthmacher, pp. 145–46; Freidel, pp. 422–30; Schlesinger, *Coming of the New Deal*, pp. 96–97; Bernstein, *Turbulent Years*, pp. 27–31; Burns, pp. 215–16; Glen Lawhon Parker, *The Coal Industry: A Study in Social Control* (Washington: American Council on Public Affairs, 1940), pp. 96–107; Lauck to Flynn, April 27, 1936, Lauck Papers; Himmelberg, pp. 190–92.

not prove easy, the trade association spokesman carried the day; the bill focused on the Chamber of Commerce plan of business self-regulation through codes of fair practice. On May 10 the two groups brought their differences to Roosevelt. He told a committee of Johnson, Perkins, Wagner, Richberg, Dickinson, and Lewis Douglas to lock themselves in a room and not come out until they had a single draft.[19]

The resultant bill was both a compromise and a composite. The provisions empowering trade associations to draft codes of fair competition—the exclusive concern of Johnson and central to both Dickinson and the Chamber of Commerce— came through intact. The licensing provision pleased the planners; the protections for collective bargaining and wages and hours kept in by Senator Wagner despite the objections of the National Association of Manufacturers had labor's support, and the public works appropriation totaling $3.3 billion, although small, was the best the public works advocates could expect from a budget-conscious Roosevelt. The president had paved the way for the measure some days before in his second fireside chat on May 7, 1933. He asserted that "it is wholly wrong to call the measures that we have taken Government control. . . . It is rather a partnership between Government and farming and industry and transportation, not partnership in profits, for the profits still go to the citizens, but rather a partnership in planning. . . ." Earlier, Moley had asked him whether he realized the implications in "taking this enormous step away from the philosophy of equalitarianism and laissez faire."

"F.D.R.," wrote Moley later, "looked graver than he had been at any moment since the night before his inauguration. 'If that philosophy hadn't proved to be bankrupt,' said the president after a period of silence, 'Herbert Hoover would be sitting here right now. I never felt surer of anything in

19. Hawley, pp. 24–25; Schlesinger, *Coming of the New Deal*, pp. 96–98; Freidel, pp 422–25; Galambos, chapter 8; Perkins, pp. 197–200; Donald R. Richberg, *The Rainbow* (Garden City: Doubleday, Doran & Co., 1936), pp. 106–107; Leverett S. Lyon et al., *The National Recovery Administration: An Analysis and Appraisal* (Washington, D.C.: The Brookings Institution, 1935), pp. 3–7; John M. Clark, *Social Control of Business* (New York: McGraw-Hill Book Co., Inc., 1939), p. 433.

my life than I do of the soundness of this passage.' " Both overestimated the impact of NIRA.

The NIRA was indeed a dramatic innovation. But it was flawed to the core economically. The public works section was woefully inadequate as an economic stimulant. Without massive federal spending, the economy would not expand to provide the jobs and production needed to bring recovery. Raising prices through trade associations and encouraging union development and eventual wage increases was no long-term solution. Like the proposals for the coal industry, the NIRA represented a compromise between labor and trade association spokesmen. The compromise insured defeat for the advocates of federal spending. Raising wages and prices would, the New Dealers later learned, not substitute for increasing aggregate demand through expansionist fiscal policy.[20]

The bill passed the House 325–76 within days and was ratified by the Senate on June 13 by a vote of 47–39 over strong opposition from antitrust progressives who feared monopolies and price fixing. It began the nation's second great experiment in business-government cooperation. "It is," said the president, "a challenge to industry which has long insisted that, given the right to act in unison, it could do much for the general good which has hitherto been unlawful. From today it has that right." Wagner, perhaps more in hope than in certainty, asserted that "it will be a powerful factor in bringing order and health into the economic life of the American people." The enactment of the law, said Harriman, was the beginning of a "new business dispensation" and marked a new rule of "constructive cooperation." Recently appointed to head the new National Recovery Administration, Hugh Johnson opened up shop even before final passage and began to recruit a staff to carry out the "charter of industrial self-government."[21]

20. Freidel, p. 424; Roosevelt quoted by Raymond Moley, *After Seven Years* (New York: Harper and Brothers 1939), p. 189, chapter 6; Roosevelt, vol. 2, p. 164. On the New Deal's failure to employ fiscal policy, see E. Cary Brown, "Fiscal Policy in the Thirties: A Reappraisal," *American Economic Review* (December, 1956): 865–66 and passim.

21. Schlesinger, *Coming of the New Deal*, pp. 99–101; Leuchtenburg, *Franklin D. Roosevelt and the New Deal*, p. 58; Freidel, p. 450; John Perry Miller, *Unfair*

The legislature passed a record fifteen major bills and adjourned the day following the signing of the law. That same evening, exhausted from the grueling legislative session, the president took a dip in the White House pool and boarded a special train "for salt water." He left the difficult task of organizing the NRA to the irascible General Johnson, backed up by a Special Industrial Recovery Board made up of cabinet members and modeled on Wilson's Council of National Defense. A blunt egotist with a mercurial disposition, Johnson realized the gravity of his assignment—and perhaps his own shortcomings, which time would reveal. This assignment, he grumbled, "is just like mounting the guillotine on the infinitesimal gamble that the ax won't work."[22]

For a spell that summer, Johnson's dynamic and excessive enthusiasm made him the central figure in Washington. "Hugh and Cry" Johnson or "Crack-down" Johnson seemed, despite his limp, to be "always in movement, even when seated." In a plain office containing some dilapidated chairs he worked long into the night, attempting to bring America's industries under their codes of fair competition. He swore, he worked, he smoked, he blustered, and he drank. His square face, marked with lines and pouches beneath the eyes, revealed the enormous strains he bore. Unable to delegate effectively, he increasingly assumed more responsibility than he could handle, then disappeared on benders. Beneath his gruff and caustic exterior was a highly nervous, sentimental person who wept at operatic arias and believed that West Point and the First Cavalry had "feelings" which he dared not hurt. His First World War work on the draft and the WIB gave him much necessary experience, but he was, as Baruch

Competition: A Study in Criteria for the Control of Trade Practices (Cambridge, Mass.: Harvard University Press, 1941), pp. 310–13; Roos, p. 52; Huthmacher, pp. 148–51; Bernard Bellush, *The Failure of the NRA* (New York: W. W. Norton, 1975), pp. 19–25; Harriman quoted by Hawley, p. 13; John D. Flynn, "Whose Child is the NRA?" *Harper's* 169 (September, 1934): 391; *New York Times*, May 19, 1933.

22. Franklin D. Roosevelt, *On Our Way* (New York: The John Day Company, 1934), p. 107; Freidel, p. 453; Roosevelt, *Public Papers*, vol. 2, p. 247; Johnson, p. 208.

said, "a good number three man, maybe a number two man, but he's not a number one man."[23]

Yet cajoling industrial managers into agreeing to codes of fair competition which would protect collective bargaining and outlaw unfair trade practices required Johnson to be at least a number one man. Code drafting proved to be a demanding, exhausting process. By mid-July only cotton textiles of the ten major industries had written a code. Alexander Sachs, brought in to head the NRA Research and Planning Division, convinced Johnson that he could not win a court test on NRA if he attempted to force codes, so the general tried voluntary action. Moreover, Roosevelt had given control of the public works division of NIRA to the parsimonious Ickes. Deprived of control over the NIRA's tool for economic invigoration, Johnson turned to moral suasion.[24]

Sometime in July the former draft mobilizer determined to launch NRA as a mass movement. Over SIRB reluctance, Johnson shifted into high gear. He sent his own sketch of the mythological thunderbird to artist Charles T. Coiner of Philadelphia. Coiner created a symbol for NRA—an eagle with a cogwheel in one claw and three lightning bolts in the other, atop the patriotic motto, "We Do Our Part." Like the insignia used by the WIB in encouraging the shoe industry to comply with WIB restrictions in 1918, the bird was sent to any employer who agreed to sign up under the "blanket code" or President's Reemployment Agreement. Mailed out to millions of employers, this agreement to work a forty-hour week, pay a minimum wage of twelve or thirteen dollars a week, and eliminate child labor became the focus for the greatest propaganda campaign since World War I. The crusty general warned slackers, "May God have

23. *New York Times,* July 23, 1933; Russell Owen, "General Johnson Wages Peace-Time War," *New York Times Magazine,* July 30, 1933; Bernard M. Baruch, *Baruch: The Public Years* (New York: Pocket Books, 1962, originally published by Holt, Rinehart and Winston, 1960), pp. 197, 207, 233; Bernstein, *Turbulent Years,* pp. 43–44; Turner Catledge, "The Fighter Who Directs Our Industry," *New York Times Magazine,* June 25, 1933, p. 3; Johnson, p. 34; Perkins, pp. 200–201.

24. Johnson, pp. 223–25; Ickes, pp. 53–54; Schlesinger, *Coming of the New Deal,* pp. 180–89; Hawley, fn. 6, p. 57; Galambos, chapter 9.

mercy on the man or group of men who trifle with this bird."[25]

"When Mr. [William Randolph] Hearst looks at that bird," wrote Arthur Brisbane, "he has to postpone his dinner." The *Daily Worker* cartooned the eagle turning into a swastika. Yet two million American employers signed the pledge. Tin Pan Alley produced a new pop tune, "Nira," and Postmaster General James Farley brought out an NRA stamp. On Wednesday, September 13, 1933, an estimated two million New Yorkers watched a quarter of a million marchers parade for NRA. Al Jolson, Ruby Keeler, and the beauties in the Radio City Music Hall Rockettes promenaded along with barbers, businessmen, West Point cadets, and Chinese school children. Carrying banners stating that they were "100 Percent Organized," marchers tramped to the strains of "Happy Days Are Here Again." Those who quit the parade to see a movie found the endless procession still marching when they came out. Lasting until after midnight, Johnson's great ballyhoo was, wrote the *New York Herald Tribune,* "New York's greatest demonstration."[26]

The NRA propaganda crusade peaked, but the economy slipped. The economic boomlet generated by industrial consumers stockpiling at pre-NRA prices collapsed in late summer. The dollar sank to a ratio of 4.86 to the British pound, causing Roosevelt to torpedo the London Economic Conference. Despite Roosevelt's willingness to release up to $30 million in gold to shore up the dollar, the stock and commodity markets plunged.[27]

Johnson used the propaganda campaign to bring the major industries under their codes during August. He personally corralled the oil men and promulgated their code. Shipbuilding, coat and suit, electrical manufacturing, and wool textile industries all signed early in the month. Roosevelt personally intervened to get the steel executives to subscribe. Automo-

25. Johnson, p. 155; Hawley, p. 53; Schlesinger, *Coming of the New Deal* pp. 112–15; Baruch, p. 68; Edward Robb Ellis, *A Nation in Torment: The Great American Depression, 1929–1939* (New York: Capricorn Books, 1971), pp. 342–43.
26. Schlesinger, *Coming of the New Deal*, pp. 115–16; Hawley, pp. 344–45; Johnson, p. 267.
27. Freidel, p. 499.

bile manufacturing—without individualistic Henry Ford—
came under its code the same day as steel. By late August,
Johnson had codified many of the basic industries. He was,
however, still struggling to crack what he called the "tough-
est nut" of them all—bituminous coal.[28]

Drafting the Coal Code

Bituminous coal proved to be difficult to codify because
the industry was a house divided against itself. Now reduced to Illinois and Indiana, who continued to operate at
the union scale, the Central Competitive Field faced stiff
competition from the non-union mines in the Appalachian
chain. Throughout the industry, wherever two fields with
different coals, freight rates, mining conditions, and wage
scales shipped to the same markets, internecine economic
warfare raged. Age-old rivalries between the unions (usually
the UMW) and the anti-union operators made bringing the
factions together on a single code of fair competition a her-
culean, if not impossible, task.

Johnson worked throughout the hot summer and into September to coerce the industry to sign the coal code, called
"the keystone to the arch of the recovery program" by the
New York Times. He had to pull "every conceivable wire"
before he could draw the warring divisions together. Getting the operators to agree "to anything at all," claimed the
New Republic, was "a triumph." The resultant code, which
Johnson called "the greatest achievement of the Recovery
Administration," like a multinational treaty, was a fragile
arrangement.[29]

The negotiations began in early June at the Washington
offices of the NCA. Following a meeting of the National

28. NRA, Committee of Industrial Analysis, "Report," (lithographed), Washington, 1937, p. 25; *New York Times,* August 3, 1933; Schlesinger, *Coming of the New
Deal,* pp. 116–17; Hawley, p. 55–56.

29. *New York Times,* June 18, 1933, section II; July 10, 11, 1933; Johnson, p.
243; Richberg to Ernest Gruening, September 16, 1933, Box 1, Richberg Papers;
New Republic 126 (September 27, 1933): 169; *Commercial and Financial Chronicle* 137 (September 16, 1933): 2037; *Business Week,* August 19, 1933, p. 5; SIRB,
"Proceedings," October 9, 1933, p. 10.

Association of Manufacturers (NAM) where coal operators had joined other industrialists in opposing the pending recovery bill because of its labor provisions, Johnson met with thirteen NCA leaders, including Francis and O'Neill, to encourage them to begin drafting a code. Two days later Johnson addressed more than one hundred operators in town for the NAM meetings. This group appointed Charles O'Neill to head a drafting committee empowered only to prepare "an outline of a code to serve as a guide to all producing fields that decided to submit a code to the Government." The committee drafted the "model" code within a week and began distributing it.[30]

Johnson realized that because this "outline" or model code omitted any mention of collective bargaining, hours, prices, or wages, it could not serve as a serious proposal. He so informed O'Neill and departed on June 16 for Chicago to address the annual National Coal Association convention. Held up in Pittsburgh by bad weather, the indomitable Johnson took to the radio waves and exhorted the operators at the Drake Hotel to organize the industry under a broad, meaningful code. A proposal from the floor of the NCA meeting that the model code include substantive provisions for hours, wages, and prices failed. On the motion of James D. Francis, the convention overwhelmingly endorsed the general model code as drafted by the special committee.[31]

Although the president could impose a code if an industry refused to draft one, Johnson relied initially on cooperation rather than coercion. But when the NCA ratified the model code, the general lost the crucial test of wills between the industry and the government. Instead of requiring the national trade association to draft a unified, industry-wide code, he had allowed the NCA to inform the regions that they should all produce their own codes. Southwestern op-

30. NRA, "Economic Survey of the Bituminous Coal Industry under Free Competition and Code Regulation," by F. E. Berquist et al., NRA Work Materials No. 69 (mimeographed), 2 vols., (Washington, D.C.: Government Printing Office, 1936), vol. 1, p. 78; Parker, p. 106; Charles Hosford testimony in *To Create a Commission*, p. 127.

31. SIRB, "Proceedings," June 16, 1933, pp. 11–13; *New York Times*, June 17, 1933; *Coal Age* 38 (July, 1933): 233; Parker, p. 107.

erators met at Denver, three Indiana operator groups convened at Terre Haute, and Alabama operators met in Birmingham to draft a code which would protect their low
wages and non-union status. Western Kentucky coal men
did the same. John Brophy urged some Pennsylvania operators to draft a code in cooperation with UMW Local 1386.
The Littleton Colorado Coal Operators Association broke
with other Colorado operators and, claiming "differences in
situation, mining operation, and coal veins," wrote a code
covering only thirty-five employees.[32] Asked by a newsman
whether he felt he could get the industry under a single
code, Johnson noted that a coal code "will have to have
variations according to regional differences." When the regional codes came in, he promised "to get together just as
large a bloc as we can."[33]

A unionized, northern operator group proclaimed itself
wholeheartedly behind the NRA and wrote a code which
dramatically reduced hours and recognized the UMW. Ben
Grey of New York City and Frank Taplin of Cleveland
formed Central Coals Associates in June and eventually
drafted a code to cover Ohio, Maryland, Pennsylvania, West
Virginia, and eastern Kentucky. A Republican friend of brain
truster Raymond Moley, Taplin worked for Roosevelt's election and had Moley's support in pressing Johnson to accept
the code. Grey worked closely with Alexander Sachs, Johnson's technical advisor. But realizing that Grey's readiness to
support a reduced work week would enrage the southern
operators, the general rejected their code, charging that Grey
was attempting to start a fight within the industry.[34]

General Johnson perhaps did not then perceive the role an
aroused union might play in the drafting process. But when
the struggle was over, he admiringly wrote Lewis that he had

32. NRA, "Economic Survey," vol. 1, pp. 78–82; John Brophy, Oral History
Memoir, p. 504, copy in Brophy Papers.
33. NRA, "Administrative Staff Studies," Records of the Division of Industrial
Economics, Staff Studies, p. 156, RG 9, NA; Johnson at press conference, July 7,
1933; transcript of conference, Bituminous Coal, General, Lauck Papers.
34. Ben Grey to Stephen Early, September 14, 1933, OF 175, Box 1, Roosevelt
Papers; *New York Times*, July 10, 14, 1933.

been "one of the greatest influences on the President's New Deal—Toleration, understanding—Statesmanship. You have no intelligent opponents on the part of employers. They regard you as a partner and so do I—but as for me—mainly . . . a friend." But even before NIRA passed Congress, Lewis had set out to insure that his union would reap the benefits of NIRA's Section 7a. Lewis launched an enormously successful union reorganization drive. By code-drafting time the avalanche of union applications flooding UMW headquarters gave the labor leader awesome bargaining power.[35]

Lewis invited every coal operator in the country to meet with UMW representatives to draft a nationwide code. Frank Taplin and other Central Field operators who either had contracts with the UMW or who hoped that a nationwide code might reduce the North-South wage differential accepted the invitation. Progressive figures in the industry like Josephine Roche, head of the Rocky Mountain Fuel Co., and others met with the union leaders on June 29 and scheduled Washington meetings to draft a national code incorporating Section 7a. Claiming to represent a "substantial proportion of the bituminous coal tonnage of the United States," these draftsmen wrote what became known as the "general code." In the last analysis it would be a truly nationwide union that would compel the operators to agree on a single code.[36]

Former Fuel Administration distribution chief James D. A. Morrow set out to meet the union challenge. He called the operators from the western part of Pennsylvania together with other non-union operators in Ohio—no longer part of the Central Field agreement—and northern West Virginia to form the Northern Coal Control Association. A key figure in bringing business talent to bear in solving the coal distribution problems of World War I, the rather dour, thin Morrow

35. Dubofsky and Van Tine quote Johnson on p. 187. The story of the revival of the UMW is in the next chapter.
36. General Code, Code Record Unit, BC, Volume A, RG 9, NA; NRA, "Economic Survey," vol. 1, pp. 79, 92; *New York Times*, July 13, 1933; Josephine Roche and Howard W. Showalter testimony, "Coal Code Hearing," transcript in Library Unit Records, Hearings, BC, pp. 2006, 3019, RG 9, NA.

wanted operators either to control the NRA or advise the NRA on coal recovery.[37]

Non-union operators in the smokeless coal region of West Virginia met at White Sulphur Springs at the end of June to form the Smokeless and Appalachian Coal Association, which combined with the Morrow group in general July meetings in Washington. Together the two groups accounted for over half the bituminous coal tonnage in America. They were led by the industry's strongest men: Morrow, James D. Francis, Charles O'Neill, and Ralph E. Taggart of Stonega Coal and Coke Company. Newsmen began calling them the "four horsemen." Wage differentials separated the northerners from their southern counterparts, but hostility to Lewis's UMW cemented the alliance. Their Appalachian code, like that for the steel industry, protected the open shop.[38]

Literally inundated with more than twenty coal codes, Kenneth Simpson, Johnson's deputy in charge of coal and a mining engineer with a degree from Columbia University's School of Mines, worked with technical advisor James H. Pierce to find the common ingredients and the representative character of the various drafts. Uniting the sponsors of the general code with those of the Appalachian draft was the key, yet not the entire objective.[39]

The UMW reorganizing drive met the determined opposition of the owners and operators of the captive mines, which produced coal for the steel industry, and compounded the problem. Both Lewis and Tom Moses, who headed the H. C. Frick Co. and represented the steel executives, realized that the captive mines were the key to organizing the steel industry, and both sides dug in for a protracted struggle. In late

37. NRA, "Economic Survey," vol 1, p. 79; Ellis, pp. 348–49; *Pittsburgh Post-Gazette*, June 1, 1933. James D. A. Morrow to Hugh Johnson, June 21, 1933; Morrow to Kenneth Simpson, July 6, 1933, both in CF, BC, General, RG 9, NA.

38. *Coal Age* 38 (July, 1933): 248–49; *United Mine Workers Journal*, August 1, 1933, p. 3; *New York Times*, July 18, 21, 1933; *Business Week*, September 16, 1933, p. 9. Philip N. Shettig, attorney, to Simpson, August 14, 1933; Morrow and E. C. Mahan to Johnson, July 28, 1933, both in Code Record Unit, BC, vol. A, part 1, RG 9, NA.

39. J. H. Pierce to Simpson, August 4, 1933, Code Record Unit, BC, vol. A, RG 9, NA.

July the campaign turned violent following an altercation between UMW sympathizers and deputy sheriffs at the mines of the H. C. Frick Co. A wildcat strike erupted and quickly spread throughout western Pennsylvania. The strikers shut down the Frick mines, but only after bloodshed. Pennsylvania's Governor Gifford Pinchot, Johnson, NRA assistant administrator Edward F. McGrady, and various industrial figures finally achieved a temporary settlement on August 4 so that the code negotiations could go forward.[40]

The commercial mines were to reopen to allow all men to return to work without discrimination against their strike participation or union affiliation, to guarantee a checkweighman, and—until a code was drafted—to settle all future disputes through a mediation board appointed by the president. In a separate letter to Johnson which omitted any mention of the UMW, Moses agreed to the same conditions for the striking miners as had the commercial mine operators. Pinchot and Lewis signed a statement agreeing to call off the strike if the conditions in the other accords were carried out. Johnson flew to Poughkeepsie, where he announced the truce, and then drove to Hyde Park for the president's final approval, which came at midnight.[41]

When they heard of the separate pacts, the captive miners balked. District 4 miners wired the president, "We have not decided about our strike yet. Our men will look the situation over and give you a definite decision by Friday of next week." When a *New York Daily News* photographer tried to get a group of miners to pose with a keg of beer, they retorted: "This is no goddamn beer party. We'll give you a strikers' picture, that's all we'll give you."[42]

To secure peace in the captive strike, Roosevelt sent

40. Edward F. McGrady to Simpson, August 1, 1933, CF, BC, Division I, Labor, RG 9, NA. Stephen Rauschenbush to Gifford Pinchot, July 27, 1933; Pinchot to Sheriff Hackney, Fayette County, July 28, 1933; Thomas Moses, president, H. C. Frick Co., to Pinchot, July 31, 1933, all in Box 2554, Pinchot Papers; *New York Times*, August 4, 1933; *Time*, August 14, 1933, p. 11; Johnson, p. 314.

41. *New York Times*, August 5, 1933; *Coal Age* 38 (August, 1933): 286; *Time*, August 14, 1933, p. 11.

42. Muriel Sheppard, *Cloud by Day* (Chapel Hill: University of North Carolina Press, 1947), p. 136; Daniel Allen, "Mine War in Pennsylvania," *Nation* 137 (August 16, 1933): 176–77; SIRB, "Proceedings," August 7, 1933, pp. 3–4.

McGrady as his personal representative to Uniontown to meet with the strikers. McGrady's pledge that "the President guarantees that there will be an end to company unions—that the Coal Code will bar them" turned the tide. As Pinchot wrote Roosevelt, "These people believe in you. . . . They trust you and they all believe that you are working to get them recognition of the United Mine Workers of America." Under the guarantees, the miners voted to return to work.[43]

With the captive strike temporarily halted, the coal code hearings began on August 9, 1933. Presided over by NRA General Counsel Donald Richberg and Kenneth Simpson, they followed the procedures Richberg had quickly established for the cotton textile industry: there was no cross-examination, but only questions from the presiding NRA officials on participants' statements. Thus, the hearings turned into a parade of statements defending localism. Western Kentucky refused to "go under the Illinois code or any other." The Appalachian group felt entitled to "a code independent and apart from any code or codes that may be formulated for other sections of the industry. . . ." Code drafting in oligopolistic industries may have worked toward cartel structures, but in bituminous coal each region wanted to maintain its competitive independence.[44]

Because wages comprised as much as two-thirds of variable costs, the wage differentials became the "pivotal issue in the code." Daily wage scales for skilled labor working inside the mine ranged from $5.80 in the Rocky Mountain area to $5.00 in Illinois and Indiana (all under union contracts) but fell to $2.40 per day in non-union Alabama. Where mines with different scales competed in the Great Lakes trade, the tidewater ports, or the industrial areas of the Midwest, wrangles ensued.[45]

43. Bernstein, *Turbulent Years*, p. 51; *Time*, August 21, 1933, p. 9; *New York Times*, August 9, 19, 1933; Daniel Allen, "Strike Truce in Pennsylvania," *Nation* 137 (August 23, 1933): 206; Sheppard, p. 136.

44. Donald Richberg, *My Hero: The Indiscreet Memoirs of an Eventful but Unheroic Life* (New York: G. P. Putnam's Sons, 1954), pp. 167–69; C. F. Richardson, president, Western Kentucky Coal Association, and Charles O'Neill testimony in "Coal Code Hearing," pp. 2110, 2133; Hawley, p. 57; Galambos, chapter 9.

45. NRA, "Economic Survey," vol. 2, p. 379; A. T. Shurick, "Coal's Prospects under the NRA Code," *Mining and Metallurgy* 14 (October, 1933): 416–21.

Pennsylvanian Charles Hosford wanted to abolish the "artificial" wage differentials favoring other areas, which were, he thought, the result of a "definite secret understanding" between northern and southern Appalachian operators. Spokesmen for the general code were willing to protect the high wages of the Far West, but they wanted to cut the North-South differential to 5 percent. Reducing the Alabama differential, retorted Birmingham's Forney Johnston, "would mean annihilation of the industry. . . ."[46] Although they included Section 7a in each code, the operators devised many modifications of the NIRA's protection for collective bargaining. Referring to the weakening of Section 7a in the automobile code, the Appalachian code said that Section 7a "shall apply to each employer in his relations to his own employees, but no employer shall be required to deal jointly with other employers, or with representatives of any employees other than his own. . . ." Alabama's code encouraged company unions. *Coal Age* demanded "clarification of phraseology" of Section 7a. Johnson said that he could not "take sides" between the closed shop and the open shop. Despite the insistence of Richberg, some operators left the hearings feeling they could interpret Section 7a to suit themselves.[47]

On unfair trade practices, codes followed the section in the National Coal Association "model" which prohibited sales below the "fair market value" of a particular type of coal (as determined by a marketing agency like Appalachian Coals, Inc.); consignment shipping; secret rebates; predating or postdating invoices or contracts; attempts to "purchase business or obtain information about a competitor by gifts or bribes"; intentional misrepresentation of analyses or sizes or of the business policy of a competitor, his product,

46. Hosford to Johnson, August 15, 1933, Code Record Unit, BC, vol. A, part 1, RG 69, NA; Hosford and Johnson testimony in "Coal Code Hearings," pp. 440–44, 2042; NRA, "Economic Survey," vol. 2, pp. 376–79.

47. Johnson in SIRB, "Proceedings," September 11, 1933, p. 24; *Coal Age* 38 (June, 1933): 173; *Princeton* (West Virginia) *Observer*, September 7, 1933; *Charleston* (West Virginia) *Gazette*, August 12, 19, 1933. Johnson stated that the open-shop clause slipped into the auto code in an "unguarded minute." *New York Times*, September 7, 1933; see also *New York Times*, August 28, 1933; Forney Johnston testimony, "Coal Code Hearing," p. 2062; *Fortune*, October, 1933, p. 57; NRA, "Economic Survey," vol. 1, pp. 85–86.

price, or financial, business, or personal standing; unauthorized use of trade names, trademarks, slogans, or advertising matter already adopted by a competitor; and inducing or attempting to induce a breach of contract between a competitor and his customer.[48]

Most codes envisioned no centralized national code administration but expected the local trade association or mining institute to act as an overseer. The general code proposed a national industrial board, but of five seats, only one went to a governmental official. The general code established boundaries for regional administrative areas but did not provide any administrative machinery. Typical of most operators, those from Appalachia agreed to meet NRA on "an equal footing, but not in a supervisory capacity."[49]

The hearings merely previewed the closed-door discussions to which Simpson told the operators to report a week later—August 22. The Pennsylvania miners, who had halted their wildcat strike to allow the hearings to proceed, saw this delay as evidence of intrigue. Governor Pinchot informed Roosevelt of potential violence in the captive mines. To the president, NRA officials repeated that the situation was "as dangerous as dynamite." Roosevelt advanced the secret meetings to August 18 and summoned Morrow, Taggart, O'Neill, and Francis. If they could not draft one code, said the president, he would draft one for them. (Pinchot had written Roosevelt the day before to "keep in mind how T.R. handled divine right Baer in 1902. I am ready any time you are.") Roosevelt, said Johnson, "softened" the operators up a little.[50]

Although the coal code was one of the most difficult to formulate, Johnson made the problem even worse. Instead of forcing operators to meet together and begin compromising their differences, Johnson conferred with them individually. Convinced that the general was all bluff, coal men

48. *Coal Age* 38 (September, 1933): 293, 314–18; "Coal Code Hearings," p. 33; Appalachian Code, Code Record Unit, BC, vol. A., R 69, NA.

49. NRA, "Economic Survey," vol. 1, pp. 98 ff., 848.

50. *New York Times*, August 16, 17, 18, 23, 1933; *Coal Age* 38 (September, 1933): 316; SIRB, "Proceedings," August 21, 1933, p. 4; Pinchot to Harry A. Slattery, August 18, 1933, Letters, Slattery Papers; *Business Week*, August 19, 1933, p. 5.

dragged their feet, stockpiled coal at precode wages, and played for time. In mid-August, just when men close to the negotiations reported that the southern operators were becoming reconciled to collective bargaining, Johnson upset the agreement by announcing that he was going to "clarify" Section 7a. The general knew little about the industry, spent only fifteen minutes at the open hearings, and lacked the steadiness necessary for sustained negotiations.[51]

Finally, Roosevelt personally succeeded in getting the operators to begin wage negotiations separately from the code discussions. Because these negotiations included both northern and southern operators for the first time in bituminous history, wage differentials would be determined by collective bargaining. When the wage negotiations began to snag, Roosevelt moved in again, called the conferees to the White House, and smoothed the way toward agreement.[52]

On September 7, without any advance notice and with only two days to file disclaimers, Johnson abruptly presented the industry with a national code. He had earlier proposed a national code based on the five regional areas suggested in the general code drafted by the UMW and the union operators, and Roosevelt had also warned the industry that he would act if the operators had not found agreement by August 29. Still, Johnson's sudden leadership stunned many operators and angered more.[53]

The eventual basis for agreement within the industry, the Johnson code provided for five regional divisions and fifteen wage districts, along the lines of the general code.

51. NRA, "Economic Survey," vol. 1, pp. 102–107; *New York Times*, August 24, September 19, 1933; Grey to Early, September 14, 1933, OF 175, Box 1, Roosevelt Papers.

52. Unsigned "Memorandum on the Codes for the Coal Industry," August 8, 1933, OF 175, Box 1, Roosevelt Papers; Johnson in SIRB, "Proceedings," August 21, 1933, p. 4; *New York Times*, August 24, 25, 1933.

53. *Coal Age* 38 (September, 1933): 316; *New York Times*, August 26, 27, September 8, 1933; NRA "Economic Survey," vol. 1, p. 97. The Central Ohio Coal Operators Association wrote Johnson urging that the NRA write the coal code. Letter, August 17, 1933, Code Record Unit, BC, vol. A. part 1, RG 9, NA. *Barrons* (September 11, 1922, p. 4) reported that after Johnson's action the owners of mines in the non-union South planned to fire workers under the president's reemployment agreement and allow them to bring suits against the companies in order to test the constitutionality of the NIRA.

Each division would establish a divisional code authority to which the president would appoint one member. A National Bituminous Coal Industrial Board of ten members—one nominated from each of the divisions and five appointed by the president—would govern. Each division would create its own labor board to handle labor disputes. Decisions could be appealed to a National Bituminous Coal Labor Board of three men: one from the miners, one from the operators, and one from the government. The maximum work week of thirty-six hours could be lengthened in peak seasons. The fair trade practices section came basically from the National Coal Association "model." Child labor, compulsory purchases at company stores, and the use of scrip were abolished. The code guaranteed the miners' right to a checkweighman and to payment on a net-ton basis.[54]

Appalachian men attacked the code as "management obliteration" and "paternalistic interference of the Administration in management." One member in ten of the national board was not enough for an area which claimed it produced 70 percent of the nation's coal. Walter Jones, secretary of the Central Pennsylvania Operators Association, sent his agency's protest, a blanket objection to the whole code, to the testy general at lunch at the Occidental Restaurant in downtown Washington. As he read the statement, Johnson fumed. He rose, stalked to the messenger, threw the paper to the floor, indicted it as an insult to the president, and refused to accept it.[55]

But the government soon bowed to the intense pressure from the industry. When the hearings resumed on September 12, Richberg told five hundred operators that the NRA's draft had been for discussion only. Richberg ignored a seven-thousand-word complaint from the Appalachian group and rebuked Alabama operators who had threatened the NRA with noncompliance. He asked that the various associations appoint two nine-man committees—one to discuss

54. *Coal Age,* 38 (September, 1933): 317; *New York Times,* September 8, 1933; NRA, "Economic Survey," vol. 1, pp. 98–99.
55. *New York Times,* September 9, 1933; *Newsweek,* September 18, 1933, p. 8; Paker, p. 109.

the substance of the code and the other to discuss administrative provisions. Under the threat of a strike of sixty thousand Pennsylvania miners, and amid great behind-the-scenes maneuvering, the committees met with Johnson that afternoon. After months of acrimonious confusion and fighting, the NRA had gotten the major operator groups to begin serious negotiations.[56]

Two days later, Roosevelt once more summoned the operators and miners to his White House study and told them that if they did not agree within twenty-four hours, he would impose a code. He postponed his yachting trip to wait. One hour past the midnight deadline, haggard operators emerged from their negotiations with the NRA. They had a final draft for the entire industry. Maximum hours were increased to forty, and the minimum age from sixteen to seventeen. Divisions I and II (Appalachian and Central Field) achieved enlarged representation on the National Bituminous Coal Industrial Board, and subdivisions within divisions would be permitted. Many coal men felt they had been browbeaten into surrendering their "industrial freedom," and Alabama and western Kentucky refused to "surrender, sign, or settle."[57]

When he signed the code on September 18, Roosevelt made three changes suggested by Simpson. He established the right of the government under the code to demand any data from the industry the administrator might require. He also reserved the privilege to appoint three additional presidential members to the National Bituminous Coal Industrial Board to keep the government from being in a minority position. In addition, he removed the "interpretation" of Sec-

56. NRA, "Economic Survey," vol. 1, pp. 99–101; SIRB, "Proceedings," September 11, 1933, p. 21. Grey to Early, September 14, 1922, p. 21; Grey, "Survey and Recommendations of the Bituminous Coal Situation, September 9, 1933," Pinchot to Roosevelt, September 5, 12, 1933, OF 175, Box 1, Roosevelt Papers; *New York Times*, September 12, 1933.

57. *Literary Digest* 116 (September 23, 1933): 10; *United Mine Workers Journal*, September 15, 1933, p. 7; Homer Cummings to Roosevelt, September 15, 1933, OF 175, Box 1, Roosevelt Papers; NRA, "Economic Survey," vol. 1, pp. 100–107; *New York Times*, September 17, 1933; *Business Week*, September 16, 1933, p. 8; SIRB, "Proceedings," November 13, 1933, p. 37; Ickes, pp. 91–92; *Time*, September 25, 1933, p. 11.

tion 7a which Richberg and Johnson had appended to the code. The code went into effect on October 2.[58]

Many voices indicated relief. Deputy Administrator Kenneth Simpson wrote the president that the coal industry "has furnished more convincing evidence for the need for the integrating force" of the National Industrial Recovery Act "than any other industry in the nation." Felix Frankfurter telegraphed Johnson that the coal code was a "great achievement." Lewis told the press that the UMW would "make every possible contribution to make it effective." William Green said that the AFL would cooperate with the miners. Although Johnson praised the union and management harmony which had been achieved, some felt that the code had been "forced down the throats of the operators." The *Wheeling* (West Virginia) *Register* compared the accomplishment with the English Factory Act of 1833. Miners everywhere sensed the promise inherent in the code's guarantees; a group of miners at one mass meeting in Pennsylvania telegraphed Roosevelt: "As Lincoln freed the Negro of the South, so has your administration freed every man and woman who toils in industry.... You have delivered the miners out of the wilderness."[59]

But the miners had actually done much of the job themselves. By flocking into the union, they eroded the arguments about the open shop advanced during the code hearings by the formerly non-union operators. By late August, 1933, there were few non-union miners to protect the open-shop clauses. Then wildcat strikers in Pennsylvania put pressure on Roosevelt and Lewis to find a way to bring industrial peace. Backed

58. U.S., NRA, *Code of Fair Competition for the Bituminous Coal Industry,* 1933; Simpson to Roosevelt, September 17, 1933, (published as part of the Bituminous Coal Code).

59. Simpson to Roosevelt, September 17, 1933; Frankfurter to Johnson, September 19, 1933, CF, BC, General, RG 9, NA; *Literary Digest,* September 9, 30, 1933, p. 7 in both issues; *New York Times,* September 19, 29, 1933; *Commercial and Financial Chronicle* 137 (September 16, 1933): 2937; *Wheeling* (West Virginia) *Register,* September 18, 19, 1933. In Harlan County, Kentucky, where enthusiasm for the NRA among operators was lukewarm at best, the *Harlan Daily Enterprise* carried only a tiny article noting that a code had been signed; all terms of the code were omitted. *Harlan Daily Enterprise,* September 18, 19, 1933; *Newsweek,* October 7, 1933, p. 4; *Coal Age* 38 (October, 1933): 325; Dan Moriarty to Roosevelt, September 18, 1933, OF 290, UMW-1930, Roosevelt Papers.

by a unionized industry, Lewis drove a hard bargain which led to a nationwide labor contract which narrowed the North-South wage differential to 8 percent. This achievement removed the central stumbling blocks to operator agreement on a nationwide code. The Appalachian Agreement of 1933 included the previously covered miners in the Central Field and added some 314,000 miners in West Virginia, Ohio, Pennsylvania, Maryland, and Tennessee. It boosted wages by 50 percent.

As the capstone of the union resurgence and the struggle to draft the NRA code, the Appalachian Agreement of September 21 marked a historic turning point. John L. Lewis, accompanied by his top officials, and James D. A. Morrow, a man who had vowed never to sign with the union, entered Room 800-B in the capital's new Shoreham Hotel. Lewis and Morrow seated themselves at opposite ends of a long table. Lewis scrawled his expansive signature on the massive document and passed it down to Morrow. Morrow signed, and the two adversaries joined in a statement: "Unquestionably this agreement is the greatest in magnitude in the history of collective bargaining in the United States."[60]

Conclusions

With Roosevelt speaking for the "public interest" of industrial peace, and Lewis forcing a nationwide labor contract in the interests of the Central Field operators, his membership, and himself, the non-union operators found themselves in a corner. The story of the coal code thus was not one of a monolithic and powerful industry extracting concessions, as was the case in many other code negotiations, but of a revived union forcing a divided industry to unite. In unison with the revived UMW, the Central Field operators led the way for a single code. In essence, Roose-

---

60. The *New York Times* called the agreement "the most comprehensive labor agreement ever signed in the United States . . ." (September 22, 1933). Dissident unionists noted that the agreement lacked district ratification. See E. A. Wieck to Oscar Ameringer, February 8, 1936, Box 15, Wieck Papers; *Time*, October 2, 1933, p. 11; Ellis, p. 349.

velt and Lewis forced a fragmented industry to try industrial self-government for a second time.

Because the code incorporated Section 7a and a variety of union objectives, coal diggers could be happy, and because a code had finally been forced, outsiders might feel that the industry was on the way to solving its problems. But the code resolved neither the basic question of industrial self-government versus governmental supervision nor that of centralization versus local autonomy. By agreeing to pay minimum wages, the operators could legally set minimum prices. Yet they lacked the machinery to coordinate prices in markets served by different marketing agencies, and higher prices would weaken coal's position in the energy market. The UMW reestablished the forty-hour week, which had been won in some fields thirty years before. In abolishing child labor, stating fair trade practices, and eliminating compulsory scrip wages and the compulsory company store, the code made changes that were both dramatic and important. The code, though, was only a compromise that needed to be tested in operation. Moreover, it was a compromise built, like the NRA itself, on the incorrect economic thinking of the day that raising wages and prices would somehow increase aggregate demand.

6

The UMW and the New Deal

AS THE CODE-DRAFTING struggle made clear, only one group in the political economy of coal could claim to speak with a single voice—the United Mine Workers of America. Internally divided during the 1920's, the UMW under John L. Lewis exploited a propitious situation during 1933 to rebuild itself into the strongest union in America. Backed by his resurgent union, Lewis silenced formal rivals, took advantage of the New Deal labor boards, and won significant gains in wages, hours, and working conditions for UMW members.

As the one national force in the industry, the UMW seemed in 1934 to be capable of achieving the age-old goal of the union and the Central Field operators: the abolition of the North-South wage differential. But in the showdown over differentials, southern operators and leading southern senators convinced President Franklin D. Roosevelt to overrule NRA Administrator Hugh S. Johnson, who had supported the union cause. The South saved its differential. But the union by 1935 made some spectacular short-run economic gains.

Rebuilding the UMW

The UMW took advantage of the collective bargaining sections of the code even before Roosevelt signed it. Well before passage of the National Industrial Recovery Act, Lewis summoned his lieutenants to launch one of the most dramatic unionization campaigns in American history. That spring, organizers flooded the mine camps with handbills stating that "a great President, a great Congress and a great union have given the coal miners this glorious opportunity to become free men and live as God intended you should live in this wonderful country of ours. ORGANIZE." Sound trucks blaring the call to membership crossed the desolate mining areas. Organizers dispensed free beer and gave the union obligation to new recruits. Alabama miners sang:

> In nineteen hundred an' thirty-three
> When Mr. Roosevelt took his seat
> He said to President John L. Lewis
> In the Union we must be.[1]

And in they went. Brilliantly led by Lewis, organizers spread the message to eager recruits. Tired and depleted after its struggle merely to maintain wage levels and to take "no backward step," the UMW transformed itself almost overnight into the strongest union in America.[2]

Various forces converged to generate the UMW success. The government warmed to unionization, particularly in the Norris-LaGuardia Act of 1932 and the NIRA. John Lewis and his team of workers exploited the new government posture. They easily reorganized the previously unionized territories. In newer fields, the organizers took advantage of the coal diggers' natural sense of brotherhood, the debilitated condition of the operators, and the powerful sense among the workers that they needed the union to force improvements in wages and working conditions. The drive quadrupled the UMW membership.

Labor spokesmen also pressured the draftsmen of the NIRA. Although the collective-bargaining provisions inserted in the early drafts of the NIRA by organized labor were more pro-labor than anything Roosevelt had originally contemplated, labor advocates further strengthened them in the

---

1. "President Roosevelt's Proclamation of Emancipation for the Miners," Lauck Papers; song quoted in Arthur M. Schlesinger, Jr., *The Age of Roosevelt*, vol. 2, *The Coming of the New Deal* (Boston: Houghton Mifflin, 1958), p. 139.

2. Leo Wolman, *Ebb and Flow in Trade Unionism* (New York: National Bureau of Economic Research, 1936), p. 11 and passim. C. L. Sulzberger, *Sit Down with John Lewis* (New York: Random House, 1938), pp. 57–58.

congressional debates. AFL President William Green got sections from the Norris-LaGuardia Act guaranteeing employees freedom "from the interference, restraint, or coercion of employers of labor or their agents in organizing or electing collective bargaining representatives" inserted into the NIRA. Senator David I. Walsh of Massachusetts amended the bill to tighten the prohibitions involving yellow-dog contracts. Pro-labor forces beat back a Senate amendment endorsed by NRA chief General Hugh S. Johnson and NRA General Counsel Donald Richberg sanctioning company unions and nullifying the intent of Section 7a. As a balance to the sections of the NIRA sponsoring collective actions by trade associations, Section 7a became a matter of equity and was a central feature of the House and Senate versions which passed by substantial margins by mid-June, 1933.[3]

Because the new law failed to provide machinery to enforce the collective-bargaining provisions, it was, however, essentially "enabling legislation and nothing more." It was an opportunity and a challenge for organized labor. Johnson boasted later that the "NRA did more in six months for both organized and unorganized labor than all the labor organizations ever did. . . ." Yet Johnson and Richberg persisted in "interpreting" Section 7a in ways which weakened its protections for collective bargaining. At a press conference Johnson announced that the NRA was "not going to be used as a machine for unionizing any industry." Richberg agreed. Roosevelt showed little sympathy for the union movement. But John L. Lewis told his fellow unionists shortly after the NIRA's passage that the UMW should ride "the crest of the sentiment that has been engendered by the enactment of the Industrial Recovery Act."[4]

3. *United Mine Workers Journal*, June 15, 1933, pp. 8–9; Edward Levinson, *Labor on the March* (New York: Harper and Brothers, 1938), pp. 50–51; Elizabeth Brandeis, "Organized Labor and Protective Labor Legislation," in Milton Derber and Edwin Young, eds., *Labor and the New Deal* (Madison: University of Wisconsin Press, 1957), p. 231.

4. *New York Times*, September 1, 2, 1933; Sidney Fine, *The Automobile under the Blue Eagle* (Ann Arbor: University of Michigan Press, 1963), p. 37; Irving Bernstein, *Turbulent Years* (Boston: Houghton Mifflin, 1970), pp. 35–36; Bernstein quotes Roosevelt on p. 172 and Lewis on p. 36; *New Republic* 76 (October 25, 1933): 294–95; Schlesinger, p. 146; Hugh S. Johnson, *The Blue Eagle from Egg to Earth* (New York: Doubleday, Doran and Co., 1935), p. 362.

Called "the key to the tremendous success of the ninety-day whirlwind drive. . . ," Section 7a became a crucial ingredient in the resurgence of the UMW. The labor protections of NIRA, Leo Wolman concluded, "became that effective spur to organization which the labor movement had presumably been unable to devise for itself in the years since 1920." John Brophy recalled, "Without the intervention of 7a there was not much hope for growth in the [miners'] union." Before the NRA, another organizer told labor researcher Edward Wieck, "You could not have come up to my house [in the mine camp]. . . ."[5]

When Lewis assembled his organizers, sometime before April 1, 1933, he told them to move quickly "before the employers woke up to the fact that there were ways of getting around the law." He gave his old friend Philip Murray responsibility for the northern districts and generalship over veterans Pat Fagan, James Mark, Lee Hall, and more than one hundred other paid and volunteer workers. To the South, Lewis sent Van A. Bittner (West Virginia); William Mitch (Alabama); and Sam Caddy, Sam Pascoe, William Turnblazer, and Ed Morgan (Kentucky). Murray spoke to four thousand at a May, 1933, mass meeting in Clymer, Pennsylvania. He recruited 693 new members and shortly afterward organized the mines owned by the Delano side of the Roosevelt family. By the official kickoff day of April 1, 1933–the anniversary of the eight-hour day in the coal industry—the UMW had laid the groundwork for the drive.[6]

The organizers blitzed the operators. They plastered mine tipples with signs and shouted that "the President wants you to join the union." Word that "John L. Lewis was having beer

5. Edward A. Wieck interview with W. A. Patton, vice-president, UMW District 5, April 27, 1934, Binder 1, Box 14, Wieck Papers; Wieck interview with William Leffingwell, unemployed miner, February 5, 1934, Binder 2, Box 14, Wieck Papers; Wolman, p. 46; Emanuel Stein et al., *Labor and the New Deal* (New York: F. S. Crofts & Co., 1934), p. 17; Brophy, Oral History Memoir, chapter 16, Brophy Papers.

6. Powers Hapgood, a sometime Lewis supporter, quoted in the *Breeze-Courier*, December 14, 1933, in box for 1934–36, Brophy Papers; *New York Times*, September 23, 1933; John Ghizzoni, UMW board member, District 2, to Lewis, June 12, 1933, box for 1930–33, Brophy Papers; *Johnstown* (Pennsylvania) *Daily Tribune*, May 15, 1933.

and sauerkraut with President Roosevelt every night, and to hell with the company guards" raced through the camps. In three months the major part of the rebuilding job was over. Within a year the UMW claimed over half a million members. In the previously non-union South, the UMW had created more than seven hundred locals. On a hot July 23, Van Bittner addressed miners in the steaming Charleston armory. Many had ridden all night in trucks from as far away as the Hazzard field in Kentucky. "You men of McDowell, Logan, Mingo, the Winding Gulf, and Kanawha field—you are now free citizens of the United States and you enjoy every right the constitution and the laws give to citizens of the United States." He later wrote that it was "the greatest meeting of my life."[7]

Commercial operators collapsed in the face of the drive. Demoralized by years of declining profits (one operator said he felt he had "one foot on a banana peel and the other in the poorhouse") even those operators who "had fought the union vigorously in former years made no real effort to checkmate its organizers." The "common difficulties" of the depression, one witness felt, had induced "a conciliatory element in employer-employee relationships in many camps." Operators in the former union territories realized that a national union would help reduce the wage differentials which had given the South an advantage. Frank Taplin, president of the North American Coal Corporation in Ohio, argued that he personally "would much prefer to deal with the United Mine Workers than with these ruthless, price-cutting, wage-cutting operators who are a detriment to the industry." The *New York Times* wrote, "The operators offered no resistance.... Shopkeepers in the coal camps hailed the union organizers with an almost evangelical fervor, supplied them with gasoline for their shabby cars, and gave them a lift in the work of organization."[8]

7. Bernstein, p. 41; Cecil Carnes, *John L. Lewis: Leader of Labor* (New York: Robert Speller, 1936), pp. 240–41; *United Mine Workers Journal*, August 1, 1933, p. 8; John Brophy, *A Miner's Life: An Autobiography*, ed. John O. P. Hall (Madison: University of Wisconsin Press, 1964), p. 236.

8. Operator testimony in U.S., Congress, Senate, Committee on Mines and Mining, *To Create a Bituminous Coal Commission, Hearings before a Subcommittee,*

The shared attitudes of coal miners simplified the union drive. Living in single-industry towns in isolated parts of the country, relying on each other for assistance underground in their dangerous work, coal miners saw themselves as a special breed. "We had all got coal dust in our blood," a Harlan County miner noted, "and once a man gets coal dust in his blood he can't never do nothing except mine coal." Clannishly suspicious of outsiders, miners tended to see their union as "a fortress, a temple, and a social club." Lewis noted, "The public does not understand, and I think never will, that almost spiritual fealty that exists between men who go down into the dangers of the mine and work together—that fealty of understanding and brotherhood that exists in our calling to a more pronounced degree than in any other industry." A song of the camps ran:

> We will have a good local in heaven,
> Up there where the password is rest,
> Where the business is praising our Father,
> And no scabs ever mar or molest.[9]

Depression poverty intensified the miners' urge to band together to improve their conditions. Poverty in the mine fields "takes tyrannical possession of the mind and keeps alive the one absorbing topic," wrote one observer. "The

72d Cong., 1st sess., 1932, p. 676; George B. Galloway, *Industrial Planning under the Codes* (New York: Harper and Brothers, 1935), p. 169; Homer L. Morris, *The Plight of the Bituminous Coal Miner* (Philadelphia: University of Pennsylvania Press, 1934), pp. 116–17; *Business Week*, March 30, 1932, p. 13; Brophy, Oral History Memoir, p. 519, Brophy Papers; Frank Taplin to George M. Jones, June 27, 1931, in NRA, Consolidated Files on Industries Governed by Approved Codes (henceforth cited as CF), Bituminous Coal (henceforth cited as BC), Division I, Labor, RG9, NA; Wieck interview with W. C. Burke, president, UMW Local 6475, February 18, 1934, Box 14, Wieck Papers; *New York Times* quoted by Bernstein, p. 42.

9. Walton Hamilton and Helen R. Wright, *The Case of Bituminous Coal* (New York: The Macmillan Co., 1925), p. 215; J. B. S. Hardman, "John L. Lewis, Labor Leader and Man: An Interpretation," *Labor History* 1 (Winter, 1961): 7; Herman Lantz, *People of Coal Town* (New York: Columbia University Press, 1958); pp. 11–12; 125–28; Carter Goodrich, *The Miner's Freedom: A Study of the Working Life in a Changing Industry* (Boston: Marshall Jones Co., 1925), p. 170; Brophy, Oral History Memoir, pp. 100–105, Brophy Papers; Harlan County miner quoted in Harry Caudill, *Night Comes to the Cumberlands* (Boston: Little, Brown and Co., 1962), p. 170; Lewis quoted in Justin McCarthy, *A Brief History of the United Mine Workers of America* (Washington, D. C., n.d.), p. 3; song quoted by James A. Wechsler, *Labor Baron* (New York: William Morrow & Co., 1944), p. 7.

sorry rehash of grievances makes the miner, according to his fiber, either an abject petitioner for charity or a rebel." The union became an instrument of hope. An organizer wrote Lewis from previously non-union Kentucky, "The people have been so starved out that they are flocking into the Union by thousands."[10]

The UMW revival resembled, in a sense, spontaneous combustion. UMW leadership and the change in government protection for collective bargaining were necessary ingredients. Yet observers noted that "the miners organized themselves." Edward Wieck, who toured the fields interviewing miners and operators, wrote that one organizer "organized" forty-two large locals in seventeen days. "Did he organize them? At the rate of two and one-half a day he spent all his time making them a little speech and giving them the obligation." Just before the passage of NIRA, former Lewis antagonist John Brophy noted "a different feeling among the miners everywhere. They seem to feel that they are once more free men, and that makes me feel good also." He recalled that the miners "moved into the union *en masse,* so that within ninety days, the bituminous miners of the country were, for all practical purposes completely organized."[11]

The New Deal Labor Boards

In recognition of the UMW's new power, the code contained the collective bargaining provisions of Section 7a without qualification. It required payment of wages semi-monthly in cash or par check and prohibited child labor below the age of seventeen. Except for maintenance men or watchmen, no employee could be required as a condition of employment to live in a company house or trade at a company store. In exchange for the operators' local code authori-

10. Malcolm Ross, "Lifting the Coal Miner out of the Muck," *New York Times Magazine,* October 1, 1933, pp. 4–5; Morris, p. 110; Garfield Lewis to Philip Murray, June 22, 1933, quoted in Bernstein, p. 37.

11. Wieck to Oscar Ameringer, February 8, 1936, Box 15, Wieck Papers; John Ghizzoni to John L. Lewis, June 12, 1933, Brophy Papers; Brophy, Oral History Memoir, Brophy Papers; Edward A. Wieck, "The Miners' Union in the Steel Industry" (typescript), pp. 15–16, Wieck Papers.

ties, the miners got tripartite Bituminous Coal Labor Boards (BCLB's) to adjudicate controversies over hours, wages, or working conditions.[12]

The code correctly assumed that the miners were "organized or associated for collective action" and that they could establish their own grievance system. Disputes concerning hours, wages, conditions of employment, or alleged violations of code guarantees went first to a local mine conference, then to a district conference, and from there to a divisional conference "as the machinery for such conference may be established." The code required operators and miners to exert "every reasonable effort" to establish grievance machinery and to "negotiate to a conclusion" such controversies wherever possible. The five regional code divisions each had one BCLB, except the large Appalachian area, which had a northern board and a southern board. These six presidentially appointed, tripartite panels (with union, management, and impartial members), answerable theoretically only to the president, could conduct collective bargaining elections.[13]

Appeal from the divisional labor boards ran to a National Bituminous Coal Labor Board (NBCLB) composed of the eighteen members of the divisional boards. The NRA administrator could convene the national board if a controversy involved employers and employees of more than one division, if the decision of a divisional board affected operating conditions of more than one division, either directly or because of its effect upon competitive marketing, or if he thought a case involved the application of a policy affecting the public or the welfare of the industry as a whole. The NBCLB met only twice, endorsed the lower ruling in seven cases, and played an insignificant role in shaping policies.[14]

12. U.S., NRA, *Code of Fair Competition for the Bituminous Coal Industry, 1933* (Washington, D.C.: Government Printing Office, 1933); Lewis L. Lorwin and Arthur Wubnig, *Labor Relations Boards* (Washington, D.C.: The Brookings Institution, 1935), p. 272.

13. *Bituminous Coal Code*, Art. 7, Sect. 5. The Legal Division of the NRA further described the powers of the boards in a set of "Rules and Regulations for Bituminous Coal Labor Boards" sent out to the BCLB's on January 3, 1934. Records of the Bituminous Coal Labor Boards (henceforth cited as BCLB Records), Division I—South, Subject File, RG 9, NA.

14. *Bituminous Coal Code*, Art. 7, Sect. 5; Lorwin and Wubnig, pp. 428–39.

Partly because of a struggle between Hugh Johnson and the advocates of labor over the composition of the National Bituminous Coal Industrial Board, Roosevelt did not name the members of the labor boards until December, 1933. Secretary of Labor Frances Perkins and John L. Lewis wanted labor represented on the industrial board. Johnson wanted a place there for himself. Roosevelt denied both demands, but Lewis later got the industry to include labor. For seats on each of the divisional labor boards the union picked some of its leaders, although it anticipated that the labor boards would "exercise small functions and be comparatively innocuous." The employer members came from the industry's large firms. Roosevelt appointed presidential members with a variety of backgrounds: a prominent industrialist and former editor of *Coal Age*, a labor arbitrator from New York, the former head of the National Catholic Welfare Council, a southern judge, General Johnson's brother, and an Interior Department lawyer.[15]

The boards faced problems of union recognition, company unions, majority rule in union elections, and discrimination because of union activity and thus could either compromise or vindicate the principles of Section 7a. But policy decisions by other agencies set the limits within which they would operate. Before the BCLB's could protect Section 7a, for example, the national administration had to resolve the issue of union recognition in the steel industry's captive mines. Following the August, 1933, strike, NRA labor advisor Edward McGrady had promised the workers recognition. While agreeing to grant their workers conditions as good as those provided by the code for employees in the commercial mines, the steel executives refused to recognize the UMW.[16]

15. *Coal Age* 38 (December, 1933): 428–29; *United Mine Workers Journal*, December 1, 1933, pp. 10–11. The UMW attitude is quoted in Andrew Pangrace, "Preliminary Abstract of Work on the Labor Compliance Activities of the Bituminous Coal Labor Boards," in Records of the NRA Organization Studies Section, p. 10, RG 9, NA. Thomas S. Hogan, "A Short History of the Activities of the Coal Labor Boards under NRA," Consolidated Files of Typescript Studies Prepared by the Division of Review, pp. 5–6, RG 9, NA; *Coal Age* 38 (December, 1933): 428–29, 439; 39 (January, 1934): 33.
16. See chapter 5.

The captive mine executives signed a September 29 agreement with the government under Section 4b of the National Recovery Act to pay code wages. "This is approved," Roosevelt added in writing, "with the understanding that hours, wages, and working conditions throughout these mines will be in all respects similar to the conditions in other mines throughout the country." Since the UMW had won the check-off of union dues from their pay elsewhere, Johnson told UMW Vice-President Philip Murray that the check-off was included. Murray ordered the strikers back to work.[17]

Despite advice from UMW headquarters that the union was now virtually recognized, insurgents in the camps shouted down proposals to return to work. Next, the operators balked at Johnson's inclusion of the check-off in working conditions, having been privately told earlier by NRA General Counsel Donald Richberg that he felt it was not part of working conditions. Roosevelt wrote U.S. Steel Chairman Myron Taylor that the check-off definitely was part of working conditions, and on October 6, he laid down the conditions for a negotiation between the steel men and the union; if collective bargaining failed to resolve the differences, he would pass on the questions involved. At this point, Richberg stated that he had been "misunderstood" regarding the check-off.[18]

It took three more weeks to achieve agreement. In late October, Roosevelt called the heads of the fourteen largest captive operations to the White House. After a meeting on October 30, the president stated that the new National Labor Board (NLB) would hold an election in November to determine the question of union representation in the cap-

17. *New York Times*, September 30, October 1, 3, 1933; U.S., National Labor Board (henceforth cited as NLB), Hearings, January 4, 1934, Records Maintained by the Library Unit, Transcripts of Hearings, pp. 13–16, RG 9, NA; Johnson to Roosevelt, October 2, 1933, OF 175, Box 1, Roosevelt Papers; Wieck, p. 59.

18. Irving Bernstein, p. 46; "White House Statement on Conciliation of Labor Dispute in Captive Mines," in Franklin D. Roosevelt, *The Public Papers and Addresses of Franklin D. Roosevelt*, 13 vols., ed. Samuel I. Rosenman, (New York: Random House, 1938–1950), vol. 2, pp. 382–84. Telephone message transcript, Roosevelt to Taylor, October 4, 1933; Roosevelt notes of October 6 meeting, both in OF 175, Box 1, Roosevelt Papers.

tives; a formal agreement "at least as favorable" as the Appalachian settlement was guaranteed to the union chosen; the check-off had been granted.[19]

Appointed by mere announcement on August 5, 1933, the NLB included Senator Robert A. Wagner as public member; Leo Wolman, chairman of the NRA Labor Advisory Board, William Green, and John L. Lewis from labor; and Industrial Advisory Board Chairman Walter C. Teagle, G.E.'s Gerard Swope, and Louis E. Kirstein for management. It settled the Reading, Pennsylvania, hosiery strike by requiring a secret-ballot election for the workers to choose their representatives. It seemed to have established both the "Reading Formula" and itself.[20]

Application of the Reading Formula to the captive mine dispute, however, confirmed Johnson's fears of management obstruction. Delayed until the end of November, the elections became controversial in themselves. The UMW defeated the company unions by a vote of 10,122 to 4,403. But at the crucial Frick mines, owned by U.S. Steel, although the union's totals were higher than those of the company unions, the UMW had only carried seven of sixteen camps.[21]

When the union proposed to begin collective bargaining, the operators refused, and the NLB held hearings on January 4 and 8, 1934. The UMW claimed that since its international officials had been listed as "representatives" of the miners on the UMW slate, the union had won recognition. The companies argued that John L. Lewis, Philip Murray, and the other elected officials must actually serve on the local pit grievance committees in the mines where they had

19. Roosevelt to Taylor and William A. Irwin, October 19, 1933, Roosevelt Papers; *New York Times*, October 30, 1933; National Labor Board, Hearing, January 4, 1933, p. 124; "An Agreement is Reached to Settle the Labor Dispute in the Captive Mines. White House Statement," Roosevelt, vol. 2, pp. 439–40.

20. The NLB handled 1,818 cases involving 914,000 workers, settled 480 strikes, and averted 197 others within the first six months of its existence. Huthmacher, pp. 153, 160–61. NRA general counsel Donald Richberg bemoaned the inherent weaknesses in the NLB. Donald Richberg, *The Rainbow* (Garden City, N.Y.: Doubleday, Doran & Co., 1936), p. 146; Johnson, p. 209.

21. *Time*, September 25, 1933, p. 11; Bernstein, p. 40; Gifford Pinchot to Charles Schwab, October 24, 1933, Pinchot Papers; James P. Johnson, "Reorganizing the United Mine Workers of America in Pennsylvania during the New Deal," *Pennsylvania History* 37 (April, 1970): 128–29; Wieck, pp. 97–100.

won election. The companies characterized the October pact with Roosevelt—in which they had agreed to the election—as a "press release" which was not a correct "statement of the agreement arrived at with the President."[22]

In its January 19 decision the board split the difference. The UMW won the right to the check-off in the particular mines where the employees had voted in it, but the contracts would not mention the union by name. "A ridiculous compromise," charged the *New Republic*.[23] The owners finally signed contracts for all the captive mines where the UMW had won, and Murray called these compacts a "signal victory for the United Mine Workers." But the contracts turned out to be better weapons for the employers than for the miners; like most coal agreements, they contained penalties for loading dirty coal, under which Frick in March fired 103 miners, all members of the UMW. When questioned by the BCLB about his company's firing practices, Frick President Thomas Moses pleaded ignorance. A retaliatory strike called by the UMW miners at Colonial brought out the Frick deputy sheriffs, who bombed several UMW homes and union leader Martin Ryan's automobile. When an investigator sent by Governor Pinchot approached, snipers opened fire. Not until World War II would the United Mine Workers of America have a solid union in the captive mines. In the early New Deal, Section 7a could not be enforced where executives resisted.[24]

Moreover, the administration vacillated on labor policies. After supporting an amendment weakening Section 7a during the passage of the NIRA, Hugh Johnson allowed the automobile manufacturers to include a clause in their code which clarified the employer's right to "select, retain, or advance employees on the basis of individual merit, without regard to their membership or non-membership in any or-

22. NLB, Hearings, January 8, 1934, pp. 53–54, 115–19.

23. *Coal Age* 39 (February, 1934): 82; Lorwin and Wubnig, pp. 187–89; "Labor and the NRA," *New Republic* 77 (January 31, 1934): 335. On the captive vote, see Wolman, chapter 6; NLB, Hearings, January 4, 1934, pp. 83–88.

24. *New York Times*, February 21, 24, 28, 1934; Wieck, pp. 118–19, 144–49; John Carmody to Thomas Moses, April 10, 1934, BCLB Records, Division I—North, Subject File, RG 9, NA; Charles A. Madison, *American Labor Leaders*, 2d ed. (New York: Frederick Ungar Publishing Co., 1950), pp. 186–89.

ganization." This traditional prerogative of management, referred to by Donald Richberg as "harmless surplussage," provided the industrialists with "an excuse to fire anyone caught organizing or affiliating with a labor union." Roosevelt eventually had to state flatly that there would be no further modifications of Section 7a. But the administration continued to hedge on the question of majority rule. Although the NLB supported this principle in the Reading Formula, Roosevelt and Johnson bargained it away in order to settle an automobile dispute in March, 1934, and effectively weakened the NLB. Not until the following September did the administration support majority rule. The *New Republic* noted during the spring of 1934 that "the administration has had no firm labor policy."[25]

The administration's shifts in labor policy buffeted the BCLB's. Roosevelt's capitulation to the auto makers in March of 1934 encouraged the steel owners to redouble their efforts against the union in the captive fields. Once the National Labor Relations Board (NLRB) replaced the NLB in June, 1934, and moved to support majority rule in the fall of that year, intransigents in other governmental departments rejected the NLRB's position. Many coal operators took their cue from the anti-unionists within the government and broke off the 1934 negotiations with the UMW. When the BCLB in the Southwest attempted to enforce majority rule bargaining on reluctant employers, "in nearly every case" they were confronted by the assertion that "the Government itself has not decided as to what its policy in that connection will be." At an NRA Labor Advisory Board meeting near the end of the NRA, Lewis denounced Richberg's abandonment of organized labor so bitterly that Richberg excused himself from the meeting. The administration disarray rankled the members of the coal boards and was, according to one, "the most important factor" in lessening their effectiveness. To the end of their existence, the boards

25. Lorwin and Wubnig, pp. 60–67, 80–83, 229–30, 268–70; Fine, pp. 71–73, 221–30; Schlesinger, pp. 146–51, 397–401, 403; "Will Roosevelt Back Up Labor?" *New Republic* 77 (May 9, 1934): 351. On Richberg's attitudes toward organized labor, see Thomas E. Vadney, *The Wayward Liberal* (Lexington: University of Kentucky Press, 1970), chapter 7.

wished for "an intelligent and cohesive national labor relations policy."[26]

The BCLB's had their own problems as well. They lacked the power of subpoena and the ability to enforce their decisions except by turning the matter over to the NRA Compliance Division.[27] The BCLB's lacked investigators and other staff. During the first months of operation, operators argued that code wage-and-hour violations should be sent to the industry-run code authority instead of the BCLB's. Only in January, 1934, did the NRA give jurisdiction on wages and hours to the labor boards.[28] Because the NLB had originally decided a question involving the captive mines, the bituminous boards did not know if they could hear disputes arising in the captives.[29]

Financial difficulties plagued the boards for over a year. The code specified that the employers and employees who nominated the members should pay "the expenses." But did direct payment by operator associations or the union violate the boards' supposed judicial independence? The UMW would support only its own representatives. In July, 1934, the presidential member for the midwestern board telegraphed the NRA that the board had been unable to meet its June payroll and was in a "financial deadlock." Not until September, 1934, did the chief NRA clerk instruct the NRA

26. "Answers of Division II of the Bituminous Coal Labor Board to Questionnaire from the NLRB," BCLB Records, Division II—Correspondence, RG 9, NA. William Jett Lauck diary, March 6, 1935; Lauck, "Docket—United Mine Workers," March 16, 1935, Lauck Papers; Hogan, "Short History," pp. 13, 26–27, 45.

27. "Hearings before the Bituminous Coal Labor Board, Division I—South, of Charges against the Clinchfield Coal Corporation for Violation of Section 7 (a), May 25, 1934," BCLB Records, Division I—South, p. 128, RG 9, NA; Newell W. Roberts, R. C. Brown, and S. E. Burt, "History of the Code of Fair Competition for the Bituminous Coal Industry" (typescript), Records of the Division of Review, Code Histories for Industries under Approved Codes, chapter 3, RG9, NA; Charles Barnes to Van A. Bittner, July 3, 1934, BCLB Records, Division I—South, Correspondence, RG9, NA.

28. Philip Murray testimony in U.S., Congress, Senate, Committee on Interstate Commerce, *Stabilization of the Bituminous Coal Industry, Hearings before a Subcommittee*, 74th Cong., 1st sess., 1935, pp. 149–50.

29. Typescript minutes of January 17, 1934, BCLB discussion on this subject in CF, BC, Code Authority–Committee–Bituminous Coal Labor Board, p. 21, RG9, NA. The NRA legal department sent out a March 7, 1934, "Memorandum on Jurisdiction," which failed to clarify the situation. BCLB Records, Divison I—South, Subject File, RG9, NA.

to pay the costs of the boards and be repaid by union and industry contributions. Not a completely satisfactory settlement, the arrangement allowed the boards to proceed with their business.[30]

Animated by their union leaders' calls about a "new day and a new deal," miners deluged the BCLB's with over two thousand complaints. Half involved charges either of discrimination for union activity or some other violation of Section 7a, and another large group concerned infractions of code wages or hours. Besides these major categories, the cases ranged far and wide—from the failure to supply scales for checkweighing the mined coal to whether a man laying track should receive code wages. Often time was spent on disputes like Pittsburgh's case "No. 111, Paul Titus vs. Bisby and Moore Company." Mrs. Titus began the action by complaining in a letter to the NRA Compliance Council in Cleveland that her husband was not being paid code wages by his "company." The letter went on to the Pittsburgh BCLB and then to the UMW District Representative for investigation. The investigator located Mr. Titus at the mine, working with his stepfather Bisby and a man named Moore in a tiny coal pit, Bisby & Moore Company. Titus embarrassedly explained that he was being treated well by his stepfather and hoped that the whole matter would be dropped.[31]

Like the Titus case, nearly three-fourths of the complaints received were simply referred back to the union's district leader, a testimony to the importance of the UMW in enforcing the code provisions. Another large number of complaints were treated informally by the presidential member as chairman. Presidential member Charles Barnes for the Pittsburgh board visited Uniontown, Pennsylvania, to hear the complaints of the miners there, then telephoned the captive mine owners to encourage them to rectify the injustices dis-

30. *Bituminous Coal Code*, Art. 7, Sect. 5. John Carmody to Philip Murray, April 19, 1934; Carmody to Mead S. Johnson, April 17, 1934, both in BCLB Records, Division I—North, Subject File, RG9, NA. John Lapp expressed his fears in letters to Newell W. Roberts, July 19, 1934, and to Wayne P. Ellis, June 20, 1934, BCLB Records, Division II, Subject File, RG9, NA; Pangrace, p. 18.

31. "President Roosevelt's Proclamation for the Miners," Lauck Papers; Pangrace, pp. 48–53; BCLB Records, Division I—North, Case Files, RG9, NA.

covered. The boards held formal hearings and issued full written opinions in only about one hundred cases.[32]

Yet despite the casual procedures and the lack of direction from Washington, the boards stood as firm as they could in favor of majority rule and of bona fide collective bargaining. The tribunals held nine elections in all parts of the nation, required that secret ballots be used, and demanded that the companies recognize and bargain with the unit selected by the majority. In Pikes Peak, Colorado, the UMW won a December, 1934, election. When the Pikes Peak Fuel Company balked at collective bargaining, the board pressed the manager until the company recognized the officials as the collective bargaining agents for their employees, although the company refused to institute the check-off.[33]

This pattern, repeated in all divisions across the country, became the central lesson of the code experience: only where the UMW could help make the decision binding could the boards make Section 7a effective. The midwestern board requested that the White Ash Coal Company "deal with representatives of the men for the purpose of making a collective bargain." The managers refused even to answer the board's "show cause" letter. The UMW could not bring its power to bear. The western BCLB did not try to force the Pike View Company of Colorado Springs, Colorado, to bargain collectively, because "We do not believe the miner's [sic] organization in that camp could sustain itself in a test with management. . . . The attitude of this company has always been anti-Union, and they represent the strongest financial interests in that part of the state."[34]

32. Pangrace, pp. 48–53; "Case 179: Footdale Mine of the H. C. Frick Coke Company," BCLB Records, Division I—North, Case Files, RG9, NA; the form letters are in BCLB Records, Division II, Subject File, RG9, NA.

33. Roberts, Brown, and Burt, vol. 1, p. 30. S. M. Thompson to Thomas Hogan, January 3, 1935; Hogan to Pikes Peak Fuel Co., January 3, 1935; UMW District 15 President Frank Hefferly to Hogan, January 17, 1935, all in BCLB Records, Division I—North, Correspondence, RG9, NA.

34. "Decision of the Bituminous Coal Labor Board, Division II. In the Complaint against the White Ash Coal Company, Wheatland, Indiana, May 22, 1934," p. 2 and passim; Ora Gassaway to Power Brothers, owners of White Ash, November 28, 1934, both in BCLB Records, Division II, Case Files, RG9, NA; "Answers of Division V of the Bituminous Coal Labor Board to Questionnaire from the National Labor Board," BCLB Records, Division V, Correspondence, RG9, NA.

Where the union could not force company bargaining or win recognition, the coal boards were similarly powerless. Harlan County, Kentucky, operators controlled nearly all of their workers' activities. The coal diggers sought help from their BCLB in vain. Of ninety-two grievance cases sent to the coal labor boards, not one resulted in a decision. The complaints were, said the board, "not actionable."[35] Late in the code experience, the labor board for the southern Appalachian territory finally scheduled hearings in Kentucky's Bell and Harlan counties. Of the operators from Bell County summoned, only one appeared. William Turnblazer for the UMW explained why the Harlan meeting was never even held: it was "too dangerous to go in there."[36]

At Henry Ford's Fordson Coal Company mines in Kentucky, the locals had members, but no militancy, because the company paid higher than code wages. Asserting that it would not become "a part of the union machinery" through the check-off, Fordson flatly refused to sign the union wage agreement. Operator officials met with the workers' grievance committees and "announced" the terms of employment. Although Presidential Member Barnes had warned them that their policies "did not constitute collective bargaining," as required by the NIRA, the Ford officials failed to appear at a hearing.[37]

The tribunals tended to support the miner who had been discriminated against or fired for union activity, but enforcement came hard. Fired for union work, Albert Timmins, employee of the H. C. Frick Coal and Coke Company, first took his case to the company union committee, which informed him that he stood no chance of being rehired, although the

35. Summary of decisions in BCLB Records, Division I—South, Board Correspondence, RG9, NA.

36. Turnblazer's testimony in U.S., Congress, Senate, Committee on Education and Labor, *Violations of Free Speech and Labor, Hearings before a Subcommittee of the Committee on Education and Labor*, 75th Cong., 1st sess., 1937, p. 3622.

37. Fine, pp. 75–95; *Coal Age* 40 (January, 1935): 47. Charles Barnes to Wayne P. Ellis, July 9, 1934; Barnes to Bittner, January 31, 1935, both in BCLB Records, Division I—South, Correspondence, RG9, NA; *United Mine Workers Journal*, November 1, 1933, p. 8; April 1, 1934, p. 7; April 1, 1934, p. 10; "Hearings before the Bituminous Coal Labor Board, Division I—South, of Charges against the Fordson Coal Company," BCLB Records, Division of I—South, Verbatim Transcripts, pp. 3, 4, 12–15, RG9, NA.

mine was currently taking on new men. Timmins wrote President Roosevelt directly; Roosevelt turned the complaint over to the Cincinnati Coal Board. When Charles Barnes, the presidential member, telephoned Frick President Thomas Moses, all he could draw from Moses was a promise to investigate. Where the union was stronger, however, complaints resulted in rehiring and payment of regular wages for the time out of work.[38]

The experience of the BCLB's paralleled that of the other national labor boards created during the New Deal. The coal boards had a natural sensitivity to the grievances of the working man, not those of his employer. Created to mediate the problems springing from the resurgence of labor under the NRA, the New Deal labor boards all received complaints from, not against, labor and were regarded by American workers as their own courts of appeal. However, no New Deal board did more than help a union maintain its strength. Where the UMW had organized and could help force operator compliance with the NIRA, the coal boards could demand rehiring of a miner who had been fired for union activity. Where the union remained weak, neither the coal board nor any other national labor tribunal could force employers to abide by Section 7a. Only union power could force an operator to bargain collectively.[39] Whipsawed by shifts in administration labor policy, lacking staff and financing, and unable to enforce their decisions against concerted management opposition, the BCLB's could not always enforce the principles of Section 7a. Created originally as part of an administrative bargain between management and labor, the BCLB's functioned more as peacemakers than as champions of the law under which they had been created.

Wages, Hours, and the North-South Differential

The newly reorganized UMW, however, forced sweeping changes in wages, hours, and working conditions and re-

38. Pangrace, p. 38; for various cases ordering rehiring and repayment of wages, see BCLB Records, Division I—North, Case Files, and Division I—South, Case Files, RG9, NA.

39. Lorwin and Wubnig, pp. 377, 412; Murray Edelman, "New Deal Sensitivity to Labor Interests," in Derber and Young.

duced the North-South wage differential. The Appalachian Agreement of September, 1933, boosted wages in some fields by as much as 64 percent and blunted the South's competitive wage advantage. The agreement used the existing contract for Illinois and Indiana as a base for the increases elsewhere. Illinois kept a $5.00 scale for inside skilled labor, Central Pennsylvania increased wages 32.5 percent to $4.60, and Ohio jumped 46.6 percent to $4.50. Alabama advanced 64.3 percent to $3.40; southern West Virginia and the Cumberlands were up 58.3 percent to $4.20. Miners and operators in the North, long the victims of wage competition from the South, saw the changes as a revival of the "principle of competitive equality," the original concept of those midwestern operators who began the search for stabilization before the first World War.[40]

The NRA supported the UMW–Central Field cause. In an early report on the code, NRA Deputy Administrator Kenneth Simpson wrote that the narrowed differentials would "still promote competitive equilibrium." The NRA, he admitted, lacked the kind of data to set precise differentials, but he claimed that operators in the "different geographical divisions within the industry" saw the new differentials as "fair and equitable." The NRA study of the impact of the code also favored lessening the North-South differential as a means to end "constant and recurring price cutting and wage slashing." The South felt pressured by a Yankee NRA.[41] Alabama chafed under its large wage increases. Operators there either complied, hoping that the NRA would restore the differential, or violated the contract and the code. Perhaps because the NRA recognized the power of the new, national UMW, or because it wanted to equalize costs throughout the industry, it never completed its report on differentials. The industry slowly grew accustomed to its

40. *Bituminous Coal Code*, Schedule A, pp. 10–11; U.S., NRA, "Economic Survey of the Bituminous Coal Industry under Free Competition and Code Regulation" (mimeographed), by F. E. Berquist et al., NRA Work Materials No. 69, 2 vols., Washington, 1936, vol. 1, pp. 22, 319–403; *Coal Age* 39 (February, 1934): 65; 38 (January, 1933): 65–71; A. T. Shurick, "Coal's Prospects under the NRA Code," *Mining and Metallurgy* 14 (October, 1933): 415; Philip Taft, *Organized Labor in American History* (New York: Harper and Row, 1964), p. 428.
41. Simpson, "Salient Facts of the Industry," *Bituminous Coal Code*, pp. x–xiii; NRA, "Economic Survey," vol. 2, p. 305.

new nationwide wage contract. But the UMW leadership prepared to seek even better terms.[42]

By mid-winter of 1933–1934, coal miners had made substantial gains under the NRA. When added to the protection of the labor provisions of the NIRA as enforced by the union and the BCLB's, the gains in the Appalachian Agreement marked a watershed in bituminous coal industrial relations. In his review "Six Months of the NRA," Donald Richberg called coal "a spectacular example" of NRA-induced progress. But the miners wanted more. "In one thing there seems to be unanimity among the rank and file, including local union officers . . . ," one man observed. "The men are of the opinion that nothing has been accomplished except the building of a foundation. Time after time men have declared: 'From now on, we get things.' "[43]

At the national convention of the UMW January 23–31, 1934, the union established its goals: the thirty-hour week, moderate wage demands, further reduction of the differential, a headquarters in Washington, D.C., "drastic action" to compel union recognition at the captive mines, a federal tax on competing fuels, government control of coal mining, continuation of the NRA beyond the expiration date in 1935, unemployment insurance, and action against dual unions. One delegate urged the union to oppose all mechanized mining. Many cheered. Lewis leaped to his feet, shouting in protest, "You can't turn back the clock!" He held the vast audience of fifteen hundred delegates in Tomlinson Hall in his palm. The audience reversed itself and cheered the man who had rebuilt the UMW.[44]

The boisterous delegates roared approval for Roosevelt, "the only President in the lifetime of most of us who had lent a helping hand to the oppressed, impoverished mine

42. NRA, "Economic Survey," vol. 2, pp. 308–309, 392–96.

43. Donald Richberg, "Six Months of the NRA," *Harvard Business Review* 12 (January, 1934): 135, 138; *Coal Age* 39 (March, 1934): 103–14; (October, 1934): 380–83; E. A. Wieck, untitled review of union strength, Box 12, p. 8, Wieck Papers; Senate Committee on Education and Labor, *Violations*, p. 3445.

44. *Coal Age* 39 (February, 1934): 83; *New York Times*, January 25, 26, 1934; Bernard Feder, "The Collective Bargaining and the Legislative Policies of the United Mine Workers of America, 1933–1946" (Ph.D. diss., New York University, 1957), p. 145.

workers." AFL President Green suggested that "the hand of Providence" had affected the 1932 election. The men called the NIRA "the greatest charter of freedom since Lincoln's Emancipation Proclamation." Secretary of Labor Frances Perkins told them that labor must have shorter hours and adequate wages. In their "fresh brushed Sunday suits," the miners shouted, "God bless you, Miss Perkins" and "more power, Miss Perkins." In an unusual move, the National Coal Association sent peace emissary Charles B. Huntress, who told the delegates that "in union there is strength." He hoped that miners and management could "let bygones be bygones, let the dead bury the dead. . . ." Even for this maladroit remark the men applauded politely. The miners were full of hope, the UMW was riding the crest of the New Deal, and John L. Lewis was king.[45]

Lewis opened the formal Appalachian wage conference which began in the nation's capital on February 28, 1934, with the demands for the six-hour day, the five-day week, a wage increase, and further reduction of the North-South wage differential. The operators might alter differentials, admitted Charles O'Neill, of Peacock, Peale, and Kerr, Inc., but they would not grant wage increases or the shorter day. Higher wages would mean higher prices and loss of markets to competitive fuels, said James D. Francis of West Virginia's Island Creek Coal Co. and Pittsburgh Coal's James D. A. Morrow. West Virginia's smokeless coal producers refused to attend the negotiations, but Lewis met them separately.[46]

The negotiations finally fell to the industry's "four horsemen" who had hammered out the final code: Morrow, O'Neill, Francis, and Stonega Coke and Coal Company's Ralph E. Taggart. Lewis, Philip Murray, District 6 President Percy Tetlow, and District 17 President Van A. Bittner pressed for reduced differentials and shorter hours. When the UMW negotiated the end of northern West Virginia's twenty-six-cent wage differential, the area's operators stalked out

45. Transcript in UMW File; *New York Times*, January 24, 25, 27, 30, February 1, 1934; *Coal Age* 39 (February, 1934): 40.
46. *Coal Age* (March, 1934): 115; NRA, "Economic Survey," vol. 2, p. 309.

and appealed to the NRA for support. Operators and miners haggled for a month. At 1:30 in the morning the day before the strike deadline, they reached another landmark wage contract, giving the 350,000 Appalachian miners a base wage of nearly five dollars a day, a seven-hour day, and a five-day week. Some put the wage increase at $100 million. "The best agreement ever negotiated in the industry," Lewis typically crowed.[47]

At NRA wage hearings during the end of March, which ran concurrently with the negotiations and provided a forum for viewpoints not represented in the negotiations, the public glimpsed the bargains that had been made privately. Called "stage plays for publicity purposes" by one southerner, these hearings provided a government forum for union complaints about particular wage levels and for progressive pleas for the thirty-hour week. Near the end, Charles O'Neill presented the fruits of the real negotiations: Code Amendment No. 1, which established the seven-hour day and five-day week and largely eliminated the wage differentials of the deep South and western Kentucky.[48]

Supported by Ohio, western Pennsylvania, and the operators of the Cumberlands to keep Lewis from striking, Amendment No. 1 touched off an explosion. It reduced the southern wage differentials even further than the code had. Alabama, for example, bore a wage increase three times larger than that of the districts sponsoring the amendment.[49] Those areas adversely affected protested vigorously and delayed acceptance. On the evening of March 31, the day before the strike deadline, Hugh Johnson, picturing himself as saving the na-

47. *Coal Age* 39 (April, 1934): 152; *New York Times*, March 23, 28, 29, 30, 31, pril 3, 5, 1934.
48. *Coal Age* 39 (April, 1934): 150–51. Division II (Indiana, Illinois, and Iowa) remained at 1933–34 wage levels; Ohio and Pennsylvania were raised 40 cents a day; western Kentucky was up 60 cents; northern West Virginia was raised 64 cents; Alabama and neighboring districts were increased $1.20. NRA, "Economic Survey," vol. 2, pp. 309–17, 397, 405–407; *Business Week*, February 24, 1934, p. 20; affidavit of William Mitch, president, UMW District 20, CF, BC, Division I— General, Labor, RG9, NA.
49. *Business Week*, April 21, 1934, pp. 14–15; *New York Times*, March 31, 1934; Blackwell Smith testimony, Bituminous Coal Industry, General Conferences, March 28, 1934 (typescript), Library Unit Records, Transcripts of Hearings, pp. 304–305, 351, RG9, NA.

tion from a "thriller" of a strike in "every bituminous mine in the country," declared an "emergency" and ordered the amendment adopted. He asserted that he "had to act. . . . I fixed the differentials arbitrarily."[50]

"Lynch law in Washington," screamed *Coal Age*. The Alabama Coal Operators Association blasted Johnson's order as "arbitrary and confiscatory." On April 6, Alabama commercial coal operators locked out fifteen thousand miners and got a temporary federal court injunction restraining Johnson from enforcing the new wages. When the operators attempted to reopen their mines on the old contract's terms, the miners stayed out. Southern spokesmen claimed that 98 percent of the business community opposed the change. In Birmingham later in the month the southern industrial council cheered speeches comparing the NRA to Sherman's March to the Sea. "Before it is all over," threatened one industrialist, "we may have secession."[51]

Western Kentucky and Fairmont County, West Virginia, refused to comply, and the UMW struck. Demand for revision swept "like a wave" across the Southwest. Operators protested that the wage increase would make competitive fuels attractive. At an April 9 postamendment hearing on the amendment, Alabama, southern Tennessee, Georgia, western Kentucky, and the Southwest operators attacked "military ringmaster" Johnson. Where, they asked, were the Pennsylvania and Ohio operators who had initiated the amendment, feasting on "their pound of flesh?"[52]

Patrick Hurley, a former miner and member of the UMW and secretary of war to Herbert Hoover, spoke at the hearing in defense of the southern position. Hurley attempted to trade on his former union membership. Lewis rose. "It is

50. *New York Times*, April 1, 4, 1934; Bituminous Coal Industry, General Conference, March 26, 1934, Library Unit Records, Transcript of Hearings, p. 22, RG9, NA; Hugh Johnson, p. 315.

51. *Coal Age* 39 (May, 1934): 158; *New York Times*, April 6, 7, 8, 17, 19, 1934; NRA, "Economic Survey," p. 315. L. C. Gunter, executive secretary, Southern Coal Operators Association, to Hugh Johnson, April 4, 1934; James W. Stafford to Johnson, April 6, 1934; Albert P. Bush to Roosevelt, April 15, 1934, all in CF, BC, Amendments and Modifications, RG9, NA.

52. *New York Times*, April 1, 4, 15, 1934; W. C. Shank to Wayne Ellis, May 16, 1934, CF, BC, Amendments and Modifications, RG9, NA.

always a matter of great pride to the United Mine Workers," he announced solemnly, "when one of its sons becomes great and famous, but it is a matter of profound sorrow when one of its sons betrays the union of his youth"—the room was silent—"for thirty lousy pieces of silver." Hurley reddened and had to be restrained from charging Lewis. "Strike out," said Lewis smugly, " 'for thirty lousy pieces of silver' and leave 'betrays the union of his youth.' " Hurley responded: "As long as I can remember, he has had his ample abdomen up against the pie counter of organized labor. John Lewis is a trade unionist for 'revenue only.' If the miners stopped paying him tonight, they would be deprived of his services tomorrow." The two narrowly avoided a fist fight.[53]

Lewis's rhetoric could not, however, enforce the code amendment. Criticism of it flooded into the offices of southern senators and congressmen. Alabama's Hugo Black and Arthur Capper, Kentucky's Alben Barkley, Georgia's Walter George, South Carolina's James F. Byrnes, and Mississippi's Pat Harrison all opposed the amendment. Roosevelt finally told Johnson to modify it. Johnson issued code Amendment No. 2 reestablishing the original code differentials retroactive to April 1. The president asked the striking miners to return to work. The Southwest continued to agitate, and the NRA offered further modified rates for that area. The UMW called off its strikes, and the southern operators ended their lockouts.[54]

But Amendment No. 1 had one more serious repercussion. Having previously flaunted Section 7a's provisions in a closed-shop agreement with a company union, western Kentucky operators, angry over loss of differentials, sought an injunction in the U.S. District Court in Louisville against

53. Selden Rodman, "Labor Leader No. 1," *Common Sense*, January, 1936, p. 12; *Coal Age* 39 (May, 1934): 198; Don Lohbeck, *Patrick J. Hurley* (Chicago: Henry Regnery Co., 1956), p. 145.

54. *New York Times*, April 23, 24, May 20, June 5, 1934; *Business Week*, April 21, 1934, pp. 14; Wieck interview with L. E. Young, Pittsburgh Coal Company, April 25, 1934, Box 14, Wieck Papers. Virtually all southern congressmen received protesting letters and telegrams. See the file in CF, BC, Amendments and Modifications, RG9, NA. Bituminous Coal Industry, General Conferences, April 19, 1934, Library Unit Records, Transcript of Hearings, RG9, NA; NRA, "Economic Survey," vol. 2, p. 318; *Coal Age* 39 (May, 1934): 196–99.

enforcement of all code wages in their mines. In a sweeping decision, Judge Charles I. Dawson asserted that "mining as such is not interstate commerce and the power of Congress does not extend to its regulation as such." The NRA wage regulations, he wrote, were "the boldest kind of usurpation." If Dawson's opinion was upheld by the Supreme Court, wrote *Coal Age*, and ". . . no power resides in NRA to establish minimum wages, except as organized labor may be strong enough to impose its will upon reluctant employers, the whole system of regulation envisaged by NIRA falls." The industry would revert to "the vicious competition" from which it "has so lately emerged." When the Supreme Court struck down the entire NRA in the *Schechter* decision a year later, Dawson's logic applied to the entire industry. By then, however, outside of western Kentucky, much of the industry had a full year under the amended rates and eighteen months under the nationwide contract, a historical first.[55]

Conclusions

The NRA unfortunately attempted to revive the economy by a misguided "cost-push" theory, which was doomed to failure. Under NRA, general prices rose faster than wages. The NRA's spreading the available work brought a psychological lift but scant economic benefit for the entire economy.[56] The code drafters checked the deflationary spiral, brought a momentary end to child labor, and stopped the massive decline in labor standards, but the codes as a whole did not generate business expansion. Indeed, they restricted output, raised prices, and were a drag on the economy. Only the coming of the war ended the depression.[57]

55. *Coal Age* 39 (June, 1934): 209–54; Charles Barnes to N. W. Roberts, June 30, 1934, CF. BC, Labor, Wages-Hours, RG9, NA; *Schechter Poultry Corp.* v. *United States*, 295 U.S. 495.
56. GNP data from *The Economic Reports of the President, 1971* (Washington, D.C.: Government Printing Office, 1971), p. 197; E. Cary Brown, "Fiscal Policy in the Thirties: A Reappraisal," *American Economic Review* (December, 1956): 865ff.
57. Ellis W. Hawley, *The New Deal and the Problem of Monopoly* (Princeton: Princeton University Press, 1966), p. 132; Leverett S. Lyon et al., *The National Recovery Administration: An Analysis and Appraisal* (Washington: The Brookings Institution, 1935).

Table 2. Employment and Wages in Bituminous Mining during the Great Depression.

Year	Number of Annual Average Bituminous Coal Employees[1]	Annual Percentage of Change	Total Bituminous Wages (in thousands) of dollars)[1]	Annual Percentage of Change	Index of Man-days of Employment[2]	Annual Percentage of Change
1929	458,700		574,810		100.0	
1930	440,800	−3.9	477,100	−16.9	87.4	−12.6
1931	407,800	−7.4	351,520	−26	71.4	−18.3
1932	350,000	−16.5	237,120	−32	57.9	−18.9
1933	366,500	+4.7	260,990	+10	62.4	+7.8
1934	423,400	+15.5	367,950	+40	67.2	+7.7
1935	435,300	+2.8	403,050	+9.5	69.6	+3.6
1936	447,200	+2.7	475,490	+18	82.1	+18
1937	455,500	+1.9	508,510	+6.9	83.3	+1.5
1938	397,700	−12.7	390,360	−23.2	65.1	−21.8
1939	360,500	−9.4	402,010	+2.98	73.5	+12.9

[1]Ellery B. Gordon and William Y. Webb, *Economic Standards of Government Price Control,* TNEC Monograph 32 (Washington, D.C.: Government Printing Office, 1940), p. 335.
[2]Harold Barger and S. H. Schurr, *The Mining Industries, 1889–1939: A Study of Output, Employment and Productivity,* Publication 43 (New York: National Bureau of Economic Research, 1944), p. 347.

Thus the short-term improvements in wages, employment, and working time in the coal industry during the NRA were part of a general downward movement in the national economy and in the coal industry economy between 1929 and 1939. Table 2 shows how there were 98,200 fewer employees in coal mining in 1939 than in 1929—a loss of 21 percent. Working time and total wages also fell during the period.

Within the framework of a depressed economy and a declining industry, the coal code and the UMW's exploitation of its opportunities under the NRA can be understood. As Table 2 makes clear, 85,000 more mines were at work under NRA in 1935 than in 1932—an increase of some 24 percent. The changes wrought by the UMW and the industry under the code thus fulfilled Roosevelt's objective of putting people back to work. The two nationwide agreements also provided working miners with larger pay envelopes. Total bituminous wages paid in 1935 were up $165,930, or some 70 percent over 1932, a spectacular gain. Average annual mine wages in 1932 had been $677. Under the NRA, the average miner pocketed an additional $419 per year. Thus, many miners were notably better off, despite other price increases. Donald Richberg proudly pointed to coal payrolls in explaining the benefits of NRA, but the industry's wages had small impact on the total economy.

Reducing hours lowered man-day production, and wage gains increased the cost of labor, thus encouraging strip mining and the use of machine loaders, as Table 3 shows. Mechanized cutting, already at high levels, did not change significantly. Increasing hourly costs for mine labor, however beneficial in the short run, would lead eventually to lower levels of employment in the industry.[58]

The code wage-and-hour changes, brought by a resourceful union, were thus a holding action, which spread the available work and improved the economic lot of the miner in the short run but which would eventually leave more miners unemployed. The UMW scored significant social gains: abolishing child labor, raising wages, and improving working conditions. The union alleviated terrible short-run conditions. But as Lewis was reported to have said, he was looking out for the current miners. Their sons would have to look for work in the cities.

Spreading the work and reducing man-day productivity, of course, hurt the industry economically, since it increased per-ton costs and—combined with the price increases to be discussed in the next chapter—made coal less competitive in the energy market. In accomplishing the goals set by the NRA cost-push theory, the coal industry was reducing its marketability as it improved the social conditions of its workers (see Fig. 2, p. 21). As was true for the entire NRA, social successes came at the expense of economic productivity and long-term expansion.

Although the UMW had to threaten to strike to win these significant short-run social improvements in wages, hours, and working conditions, and did strike to try to win recogni-

58. U.S., Federal Works Agency, WPA, National Research Project, *Mechanization, Employment, and Output per Man in Bituminous Coal Mines* (Washington, D.C.: Government Printing Office, 1939), p. 333. Mechanization increased most dramatically in the Midwest, improving its competitive position. Reed Moyer, *Competition in the Midwestern Coal Industry* (Cambridge, Mass.: Harvard University Press, 1964), p. 179.

Table 3. Production, Expansion, and Mechanization in Bituminous Mining in the Great Depression.

	Number of Commercial Mines[1]	Percentage of Change	Production (in thousands of tons)[1]	Percentage of Change	Percentage Cut by Machine[1]	Percentage Mechanically Loaded Underground[1]	Production per Man-day[2]
1929	6,057	−27.4	534,989	−12.6	75.4	7.1	4.85
1930	5,891	−42.2	467,526	−18.2	77.5	10.0	5.06
1931	5,642	−38.1	382,089	−18.9	79.1	12.4	5.30
1932	5,427	+2.3	309,710	+7.7	78.8	11.6	5.22
1933	5,555	+11.2	333,631	+7.7	80.0	11.3	4.78
1934	6,258	+.9	359,368	+3.6	79.2	11.5	4.40
1935	6,315	+8.8	372,373	+17.9	78.9	12.7	4.50
1936	6,875	−4.7	439,088	+1.5	79.3	15.3	4.62
1937	6,548	−11.7	445,531	−21.8	NA	18.7	4.69
1938	5,777	+.7	348,545	+13.3	79.9	24.4	4.89
1939	5,820	+8.6	394,855	+16.7	79.5	28.0	
1940	6,324	+7.9	460,772	+11.7	80.1	32.1	
1941	6,822		514,149		79.5	36.3	

[1]U.S., Bureau of the Census, *Historical Statistics of the United States: Colonial Times to 1957* (Washington, D.C.: Government Printing Office, 1960), p. 356.

[2]Ellery B. Gordon and William Y. Webb, *Economic Standards of Government Price Control*, TNEC Monograph 32 (Washington, D.C.: Government Printing Office, 1940), p. 332.

tion in the captive mines and elsewhere, the mine fields were relatively free from labor disturbance under the NRA. Once the UMW established itself in the spring and summer of 1933, it bargained collectively in relative peace. Despite their inability to enforce Section 7a where the UMW was weak, the BCLB's helped legitimize the union and resolve labor-management disputes. In other industries the NRA touched off a chain explosion of violent strikes, particularly in 1934. By comparison, the formerly strike-prone bituminous industry was quiet, functioning under a nationwide contract which had established both grievance machinery and the BCLB's. The NRA Division of Review concluded that the industrial stability in bituminous coal was "one of the outstanding achievements of operation under the code."[59]

Indeed, the bituminous industry made significant progress in labor-management relations under the NRA. Under the constant pressure of the union, the industry established a nationwide wage scale which equalized competitive conditions, increased wages, shortened hours, improved working conditions, and, most importantly in the minds of many New Dealers, put thousands of unemployed miners back to work. The union reduced the North-South differential and thus created a truly national wage scale. That the Central Field operators and the UMW could not eliminate the differential shows the political power of the South in the New Deal coalition. The union's short-run economic and social gains, however, speak to the strength of the reactivated UMW, which gave John L. Lewis the power to bargain from strength.[60]

59. U.S., Bureau of the Census, *Historical Statistics of the United States: Colonial Times to 1957* (Washington, D.C.: Government Printing Office, 1960), p. 99; NRA, "Economic Survey," vol. 1, p. 6.

60. Robinson Newcomb, *How the NRA Worked* (New York: American Council of the Institute of Pacific Relations, 1936), p. 23.

7

Price Stabilization under the Blue Eagle

MODELED AFTER THE agencies created to sponsor business-government cooperation during the mobilization for the First World War, the National Recovery Administration seemed to have much the same style, spirit, and structure. As was true for the War Industries Board, the Fuel Administration, and the Council of National Defense—with its Advisory Commission and coordinating committees of industrialists—business met government to solve a national crisis. Indeed, the NRA code authorities replaced the cooperating committees of the CND and WIB. Yet in the two eras, government had opposite objectives. War government called on operators in the coal production committee to stimulate production, improve distribution, and hold down prices. When the operators fixed prices that seemed too high in June, 1917, President Wilson intervened and cut them by one-third. Despite their reliance on the analogy of war, the New Dealers wanted to halt falling prices. Facing strong demand pressures, the Fuel Administration set maximum coal prices; facing excess supply, the NRA set minimum coal prices.[1]

1. U.S., NRA, "Economic Survey of the Bituminous Coal Industry under Free Competition and Code Regulation" (mimeographed), by F. E. Berquist et al., NRA Work Materials No. 69, 2 vols., Washington, D.C., 1936, vol. 1, p. ii; William E. Leuchtenburg, "The New Deal and the Analogue of War," in *Change and Continuity in Twentieth Century America*, ed. John Braeman et al. (New York: Harper and Row, 1964), pp. 81–144; Gerald D. Nash, "Franklin D. Roosevelt and Labor: The World War I Origins of Early New Deal Policy," *Labor History* 1 (Winter, 1960): 39–52; Gerald D. Nash, "Experiments in Industrial Mobilizations: W.I.B. and N.R.A.," *Mid-America* 45 (January, 1963): 157–74.

However shortsighted the "reflation" economics of this approach was, coal industrialists were not alone in trying to halt price declines during the depression. In 79 percent of the NRA codes, industrialists wrote trade practice provisions involving minimum prices. Two-thirds of all codes prohibited sales below cost. Open-price filing was provided for by 60 percent. Nearly three-quarters required uniform cost accounting to end sales below costs. More than 100 of 677 codes prohibited "destructive price cutting." Under the slogan of "self-government in industry," NRA codes also regulated nonprice competition, sales practices, production control, and limitations on machine or plant hours and established code authorities to enforce these and other fair trade practices.[2]

Yet although coal men knew stabilizing prices would be difficult, they wrote one of the few codes which provided for direct price fixing. Raising coal prices became a central objective of the code. "All [The industry's] problems," one UMW spokesman noted during the hearings on the 1928 Watson bill, "can be summed up and stated in one word: 'price.'" Indeed, to be able to pay the increased wages of the Appalachian Agreement of 1933, operators had to raise prices. "The protection of the price structure of the industry was," wrote the official NRA historians of the coal code, "the most essential problem that confronted the industry in all its branches. . . ."[3]

The task seemed staggering. The National Industrial Re-

2. Leverett S. Lyon et al., *The National Recovery Administration: An Analysis and Appraisal* (Washington, D.C.: The Brookings Institution, 1935), pp. 570–71, 603; Arthur M. Schlesinger, Jr., *The Age of Roosevelt*, vol. 2, *The Coming of the New Deal* (Boston: Houghton Mifflin, 1958), pp. 124–25; Ellis W. Hawley, *The New Deal and the Problem of Monopoly* (Princeton: Princeton University Press, 1966), pp. 55–62; U.S., Temporary National Economic Committee, *Competition and Monopoly in American Industry*, by Clair Wilcox, TNEC Monograph 21 (Washington, D.C.: Government Printing Office, 1940), pp. 261, 264–65.

3. John L. Lewis testimony, "Hearing Before the National Bituminous Coal Industrial Board" (typescript), January 9, 1935, Library Unit Records, Transcripts of Hearings, (henceforth cited as "NBCIB Hearing"), p. 10, RG9, NA; Hawley, pp. 57–58; K.C. Adams testimony, U.S., Congress, Senate, Committee on Mines and Mining, *To Create a Bituminous Coal Commission, Hearings before a Subcommittee*, 72d Cong., 1st sess., 1932, p. 810; Newell W. Roberts, et al., "History of the Code of Fair Competition for the Bituminous Coal Industry" (typescript), pp. 14, 42, Records of the Division of Review, Code Histories, RG9, NA.

covery Act was vague on price fixing. The coal code provided no system for establishing minimums and no mechanism for correlating minimums in markets shared by two or more of the twenty-five code regions. The NRA leadership could not decide whether to support free, competitive markets or allow fixed prices. The operators could not prosecute price chiselers or—once the initial patriotic hoopla wore off—even sustain general compliance throughout the industry. Consumers launched legal attacks on the price-fixing process. Could such a geographically divided industry make "self-government" work to achieve price stabilization?

Initial Difficulties

Drawn from diverse sources, the NIRA directed the administration to "promote the organization of industry for the purpose of cooperative action among trade groups." It proposed to end "destructive price cutting . . . to promote the fullest possible utilization of the present productive capacity of industries. . . ." But what was "destructive price cutting"? "The diversity of NIRA authorship . . . ," wrote NRA Deputy Administrator Alexander Sachs, "has produced instead of a synthesis, a conglomeration of purposes, an obfuscation of ends and a stultification of methods." Roosevelt admitted: "NRA was deliberately conceived in controversy. It was deliberately set up as a forum where views of conflicting interest can meet in the open and where, out of controversy, may come compromise."[4]

One such compromise engineered by NRA Administrator Hugh S. Johnson and Kenneth Simpson, deputy administrator for resource industries, was the coal code. Most regional coal codes submitted to NRA contained provisions for price

4. National Industrial Recovery Act, 47 Stat. 195 (1933); Sachs to Johnson, May 20, 1933, in Charles F. Roos, *NRA Economic Planning* (Bloomington, Ill.: Principia Press, 1937), pp. 46, 472; Robinson Newcomb, *How the NRA Worked* (New York: American Council Institute of Pacific Relations, 1936), p. 7; Lyon, pp. 25, 603–605, 795; Hawley, pp. 19–20; U.S., NRA, Committee of Industrial Analysis, "Report" (lithographed), Washington, D.C., 1937, p. 163; Franklin D. Roosevelt, "The Goal of the NIRA," June 16, 1933, in Franklin D. Roosevelt, *The Public Papers and Addresses of Franklin D. Roosevelt*, ed. Samuel I. Rosenman, 13 vols. (New York: Random House, 1938–50) vol. 2, 246–77.

control. Some wanted to use Appalachian Coals, Inc., and others wanted regional trade associations. Pushed aside by the intense struggle over wage differentials, the price provisions of the code received little attention during the hearings. The code bargaining cannot be reconstructed precisely. But apparently in exchange for raising wages and for uniting under one code, the operators got the right to establish "fair market prices" through regional marketing agencies or, lacking a marketing agency, through a divisional code authority.[5]

The price sections of the bituminous code were as imprecise as the NIRA itself. The code outlawed sales of coal for less than a "fair market price," but the code did not define "fair." It said only that the "fair" price had to be high enough to "furnish employment for labor" and to pay the new rates established in the Appalachian Agreement. Because of the accounting difficulties involved, the code did not require the operators to apply a strict cost system. They could take "competition with other coals" and "the various conditions and circumstances" involved in particular coal sales into account when they established prices.[6]

The code then allowed some two dozen marketing agencies or code authorities to interpret these vague standards. Worse, the code failed to provide a specific governing agency for the industry. The NIRA made no mention of anything like a "code authority," a structure which first appeared in August, 1933, in the code for the coat and suit industry. Most coal-producing areas did not have marketing agencies like Appalachian Coals, Inc., so the task of price fixing fell to untested code authorities—ad hoc groups hastily assembled during late September, 1933. Despite complaints that certain code authorities were not "truly representative of at least two-thirds of the commercial tonnage of any coal district," Johnson on October 3 approved,

5. See Records of the Code Record Unit, BC, vols. A and B, RG9, NA, for the multitude of codes offered by various sections of the industry. Newcomb, p. 29; "NBCIB Hearing," January 3, 1935, p. 18; NRA, "Economic Survey," vol. 2, pp. 500–506; NRA, Committee of Industrial Analysis, "Report," pp. 161–68.

6. *NRA Code of Fair Competition for the Bituminous Coal Industry* (Washington, D.C.: Government Printing Office, 1933), Art. 6, Sects. 1–2.

in a blanket endorsement, all twenty-five sets of code authority bylaws.[7]

Reflecting the splintered industry, the code provided no agency to coordinate prices among regional marketing associations or code authorities. Established only because the administration demanded it, the National Bituminous Coal Industrial Board (NBCIB) was the creature of the divisional code authorities, not their master. It met twice. Roosevelt amended the final code draft to allow him to appoint "not more than three" members to the NBCIB, but these presidential representatives could not control a nine-member board. Moreover, Roosevelt's appointees to the code authorities could neither vote nor compel joint action among the authorities. They could only approve or disapprove prices set by the industrialists. Of all the important codes, one presidential member felt, coal had the weakest administrative structure. Even Johnson himself had no legal mandate to compel cooperation on prices among the industrial groups.[8]

The national administration of NRA split over price-fixing. The continuing, heated dispute among supporters of price fixing and advocates of the free market disrupted the operators' attempts to raise bituminous levels. On June 22, 1933, Johnson declared that price fixing would not be permitted

7. Ibid.; "History of the National Recovery Administration," (typescript), p. 2, Records of the NRA History Unit, RG9, NA; Robert H. Connery, *The Administration of an N.R.A. Code: A Case Study of the Men's Clothing Industry* (Chicago: Social Science Research Council, Studies in Administration, No. 4, 1938), pp. 22–23; NRA, Committee of Industrial Analysis, "Report," pp. 26, 78–80; U.S., NRA, "Code Authorities and Their Part in the Administration of NIRA, NRA Work Materials No. 46, Washington, D.C., 1935, pp. 64, 82; Roberts, vol. 1, p. 4; NRA, "Economic Survey," vol. 2, p. 501. Kelly's Creek Colliery to Wayne P. Ellis, June 29, 1934, Consolidated Files on Industries Governed by Approved Codes (henceforth cited as CF), Bituminous Coal (henceforth cited as BC), Southern Subdivision, Code Authority, RG9, NA.

8. *Bituminous Coal Code*, Art. 7, Sect. 4; F. E. Berquist to Leon Henderson, March 16, 1934, CF, BC, RG9, NA; General Frank Haas, NRA technical advisor on coal, to William Emery, Jr., chairman, Ohio code authority, November 23, 1933, RG9, NA; Emery to Haas, November 21, 1933, CF, BC, Northern W. Virginia, Meetings, RG9, NA; Walter Steinbugler testimony, U.S., Congress, House, Committee on Ways and Means, *Stabilization of the Bituminous Coal Mining History, Hearings*, 74th Cong., 1st sess., 1935, p. 205; NRA, "Economic Survey," vol. 2, pp. 501–502; Glen Lawhon Parker, *The Coal Industry: A Study in Social Control* (Washington: American Council on Public Affairs, 1940), pp. 111, 114; Bernard Sternsher, *Rexford Tugwell and the New Deal* (New Brunswick: Rutgers University Press, 1964), p. 161.

under NRA. A day later he said that NRA would allow price fixing that prohibited sales for less than the cost of production. From then on, however, he defended the price-stabilizing arrangements in the codes against the attacks of Rexford Tugwell and others. At meetings of the Special Industrial Recovery Board (SIRB), Tugwell bitterly protested against allowing businessmen to dominate a price-fixing apparatus. He called Johnson's emphasis on the propaganda drive "Rotarian Hoopla" and said that the codes businessmen dominated were "bribes."[9]

Joined by Agriculture Secretary Henry A. Wallace, Tugwell kept up his litany until Johnson persuaded Roosevelt to abolish the SIRB in December, 1933. Apparently convinced of the soundness of price fixing in the natural resource industries, however, Roosevelt refused—at that time—to undermine his NRA administrator. Consumer advocate Mary Rumsey, head of the NRA Consumers' Advisory Board, left-wing members of the Department of Agriculture, and others continued the barrage against fixed prices. By delaying public hearings on price fixing, Johnson prevented an open fight among government officials. But the dispute simmered, waiting for an opportunity to erupt.[10]

It took NRA nearly a month after the October 2 effective date of the coal code to spell out a policy on code authority organization and procedure. Roosevelt did not complete his list of presidential members for the code authorities and labor boards until late November. Johnson restructured his administrative organization somewhat in late October and appointed former coal deputy Kenneth Simpson head of a new division encompassing all of the extractive industries. Fuel Administration veteran Wayne P. Ellis, Simpson's assistant, took the coal deputy position. But the general had no price policy. Operators could not get specific answers on price fixing from NRA headquarters. Well-organized industries like cotton textiles took advantage of such a fluid situa-

9. Sternsher, pp. 164–68; Schlesinger, pp. 122–24; Rexford Tugwell, *The Democratic Roosevelt* (Baltimore: Penguin Books, 1969; originally published by Doubleday & Co., 1957), p. 312–13; *New York Times,* June 23, 24, 1933.
10. Hawley, pp. 72–77; Schlesinger, pp. 49–60, 123–30; Lyon, chapter 21.

tion. But the fractured, leaderless coal industry, without a clear direction from the law, the code, or the administration, wondered how to establish its "fair minimum prices."[11]

Despite all these difficulties, coal operators and NRA officials worked out a system which raised average per-ton coal prices 37 percent, from $1.20 to $1.75—enough to cover wage increases and other variable costs (see Table 4). Although the NRA price increases did not cover capital charges, before-tax earnings in 1934 for the industry approached the break-even point for the first time since 1927. Operator-constructed price-fixing apparatus created enough monopoly control in enough markets to make minimum prices effective. Policed by a revitalized union, the wage increases of 1933 and 1934 formed a cost base which encouraged the operators temporarily to abandon price slashing. Although the competitive sellers fought numerous interregional price wars, once the industrialists in their code authorities abandoned mine prices for market-area pricing, they achieved some price stability.[12]

In one sense, market-area prices overcame the industry's regionalism during the depression as the zone plan had done during the war. But the NRA was the mirror image of the Fuel Administration. During the war, government consumed coal and increased the already enormous demand. Over operator opposition, the officials of the Wilson years fixed maximum prices. Coal men then rescued the Fuel Administration by helping streamline distribution through zones. During the depression, many coal men wanted minimum prices, but this time government had to first pull them into a single code and then help them coordinate market-area prices that highly competitive sellers would not break.

11. David C. Reays, secretary, Northern West Virginia Code Authority, telegram to Roosevelt, November 11, 1933, CF, BC, No. W. Virginia, Code Authority, Members, RG9, NA; Hugh S. Johnson to K. M. Simpson, October 6, 1933, CF, BC, Division I, Administration Members, RG9, NA; NRA Release 1845, CF, BC, Administration Members, RG 9, Hawley, pp. 63–65; *Coal Age* 38 (December, 1933): 428; 39 (February, 1934): 71; Connery, p. 123; *New York Times*, October 26, 1933; Louis F. Galambos, *Competition and Cooperation: The Emergence of a National Trade Association* (Baltimore: Johns Hopkins University Press, 1966), chapters 7–8.

12. Waldo E. Fisher, *Economic Consequences of the Seven-Hour Day and Wage Changes in the Bituminous Coal Industry* (Philadelphia: University of Pennsylvania Press, 1939), p. 18.

The war Congress produced a tough statute empowering the government in effect to take over the industry and tell it what it could charge and where it could ship. The depression legislators produced a law which only vaguely encouraged the NRA to stabilize prices and did not give the government anywhere near the power that the Fuel Administration had. If the operators were to stabilize prices under the NRA, they would have to govern themselves.

At first, therefore, NRA price fixing succeeded only moderately. Left largely on their own, operators developed a whole variety of price-fixing systems. Established on the geographical outline of traditional producing districts, code authorities whose members competed in the same markets soon began to quarrel.[13] Indiana's subdivisional code authority established zone prices and incorporated freight differentials. Illinois fixed only mine prices. Indiana began to undersell coals from Illinois, Iowa, and western Kentucky. They all protested, and Indiana retorted that its zone system was temporary and blamed NRA for failing to lead. Illinois retracted its minimums and left operators there to sell coal at any price.[14]

The code allowed code authorities to establish prices for "various consuming markets." But in November, 1933, Frank Haas, NRA technical advisor on coal, told Indiana that "zoning was out." Division Administrator Simpson thereupon issued an order allowing Indiana her market zones. When the midwestern subdivisions began to fix zone prices, the Appalachian division announced that it might be forced to "enter the warfare" if the midwestern zone prices gave the Central Field an edge in competitive markets.[15]

13. The code divided the coal-producing areas into twenty-five subdivisional code authorities.

14. Frederick Delano to Wayne P. Ellis, January 8, 1935, Delano Papers; Hawley, P. 61; Roberts vol. 1, p. 31; James D. Francis testimony, House Committee on Ways and Means, *Stabilization*, p. 434; "Record of Conference of Representatives of Coal Producers from Iowa, Illinois, Indiana, Western Kentucky, and Eastern Representatives, December 7, 1933," p. 12, CF, BC, Division I, Prices, Conferences, RG9, NA; Connery, pp. 25–27.

15. "Record of Conference, December 7, 1933," pp. 1–4, 10, 18, CF, BC, Division I, Prices, Conferences, RG9, NA; *Coal Age* 38 (December, 1933): 428–29; (November, 1933): 390–91.

In part, the difficulties of the first months reflected the industrialists' own ambivalent attitudes. They sought leadership from the NRA but wanted to keep control in their own hands. The northern West Virginia presidential member wrote that operators there would have followed a dictatorial policy from NRA in order to stabilize prices. A Peabody Coal Company spokesman pleaded with the NRA to issue flat, direct statements of policy. The industry needed, others said, "a rule and a broad policy" from NRA. But James D. Francis told the NBCIB that he feared government intervention, which might end in "giving power to government to fix prices." Protect your code authorities, urged one coal lawyer, "see that they are not shorn" of power "by some Washington bureau."[16]

Still, the code pricing systems began to work. Competition among subdivisions over the higher, fixed prices replaced competition among individual firms over lower, pre-code levels. The operators gradually created a degree of monopoly control in particular markets. The Illinois–Indiana–western Kentucky controversy required numerous meetings that failed to satisfy the disruptive western Kentucky operators. Yet the average prices for Illinois and Indiana mines from December to April were up forty-five cents—a gain of one-third over pre-code contract prices. Subdivisions in the giant Appalachian division maintained artificial price levels by competing on quality. Yet by April the Appalachian prices had risen by approximately the same percentage as those in Indiana and Illinois.[17]

By January, 1934, *Coal Age* wrote that there was a "general agreement" in the industry "that the code has laid the

16. "NBCIB Hearing," January 16, 1934, p. 24; January 19, 1934, pp. 10–31; Philip Murray testimony, U.S., Congress, Senate, Committee on Interstate Commerce, *Stabilization of the Bituminous Coal Mining Industry, Hearing before a Subcommittee,* 74th Cong., 1st sess., 1935, p. 151; George Hadesty, presidential member, memorandum on Northern W. Virginia, June 10, 1935, in Roberts, vol. 3, p. B:8; C. W. Reed, remarks, "Record of Conference, December 7, 1933," p. 23, CF, BC, Division I, Prices, Conferences, RG9, NA; E. L. Greever, general counsel, Pocahontas Coal Operators' Association, quoted in *Coal Age* 39 (January, 1934): 9.

17. "Record of Conference, December 7, 1933," pp. 10–18, CF, BC, Division I, Prices, Conferences, RG9, NA; *Coal Age* 38 (November, 1933): 390–91; (December, 1933): 428–29; 39 (February, 1934): 32, 71; (September, 1934): 364–65; NRA, "Economic Survey," vol. 2, pp. 518–19, charts I and II, appendix 3.

groundwork for stabilization of the industry on a sound economic basis. Hugh Johnson boomed, "Nowhere in the President's recovery program is there a clearer example of teamwork between government, industry, and labor" than in bituminous coal. Code authorities praised Deputy Administrator Wayne P. Ellis for the successful price program. At an early spring United States Chamber of Commerce meeting of natural resource industry representatives, the coal men were among the most enthusiastic supporters of the NRA.[18]

Consumers Retaliate

Just as coal operators congratulated themselves for raising prices, however, free-market spokesmen made their influence within NRA felt. The consumers' advisory board members, Tugwell had said, wore "spearheads without shafts." But in December, 1933, Mary Rumsey brought the forceful Leon Henderson of the Russell Sage Foundation to meet with Johnson on consumer problems. After Henderson and the general finished a brief shouting match, Johnson invited him to become an assistant on consumer problems in NRA. Soon Henderson was head of NRA research and planning and the chief economist of the Recovery Administration. Johnson thought Research and Planning was only a statistical office, but Henderson and his staff made it a staging ground from which to attack the pricing policy of NRA.[19]

Free-market advocates launched the opening volley against the NRA's price policies at Johnson's much-delayed public price hearings on January 9, 1934. Government purchasers, farm spokesmen, and other consumer advocates denounced both the increased prices brought by NRA and the

18. *Coal Age* 39 (February, 1934): 72; H. C. Marchant, member, Division 5 Code Authority, telegram to Ellis, May 15, 1934, CF, BC, Division 5, Code Authority, Reports, RG9, NA; Warren E. Pippin, executive secretary, Michigan Code Authority to Ellis, May 18, 1934, CF, BC, Division 1, Code Authority, Reports, RG9, NA; Irvin Davis, chairman, Southern Sub-Divisional Code Authority, to Ellis, May 18, 1934, CF, BC, Division 1, Code Authority, Reports, RG9, NA; Resolution of Northern W. Virginia Code Authority January 15, 1934, CF, BC, Code Authority, Meetings, RG 9, NA; Hugh Johnson, "The New Order," *American Federationist* 40 (November, 1933): 1185; *New York Times*, May 3, 1934.

19. Hawley, p. 76–79; Lyon, pp. 705–707; Schlesinger, pp. 127–32.

seeming collusion of some uniform bids. Critics singled out soft coal. When Arthur D. Whiteside of Dun and Bradstreet, who chaired the hearings, refused to accept detailed studies on prices in particular industries prepared by the Consumers' Advisory Board, the press headlined the "gagging" of the CAB. Soon the big guns of the antitrust arsenal—veteran Senate Republicans of the old progressive persuasion like North Dakota's Gerald Nye and Idaho's William E. Borah—opened fire. Nye charged that NRA programs were breeding monopoly, encouraging price fixing, and crushing small businessmen. Borah proposed an amendment revoking the NRA's suspension of the antitrust laws.[20]

In response, Johnson scheduled further, open hearings on NRA for February 27, 1934. For four days labor spokesmen, consumer advocates, small businessmen, and persons representing a whole spectrum of political points of view lambasted the NRA as monopolistic, antilabor, oppressive, and fascistic. On March 5, just over a year after the inauguration of the New Deal, Roosevelt responded to NRA critics by telling a meeting of the code authority leaders of some six hundred industries they were "the largest, most representative conference of American business ever held."[21]

Recall, he urged, the pre-NRA situation when the attitude of "every man for himself, the devil take the hindmost" prevailed. "Special groups . . . who believed in their superhuman ability to retain in their own hands the entire business and financial control over the economic and social structure of the Nation" had control of government. The NIRA, he said, was "drawn with the greatest good of the greatest number in mind." Use the "representative government in industry" program of NRA, he advocated, to "play the game." As code authority leaders, be "your industrial

<hr>

20. Roosevelt, vol. 2, p. 278; *New York Times,* December 27, 1935; Leverett S. Lyon and Victor Abramson, *The Economics of Open Price Systems* (Washington, D.C.,: The Brookings Institution, 1936), p. 25; Hawley, pp. 79–81; Schlesinger, p. 131; Lyon, pp. 41, 706–10; Herbert F. Taggart, *Minimum Prices under the NRA* (Ann Arbor: University of Michigan Press, 1936), pp. 26–28.

21. Roosevelt, vol. 3, pp. 131–32, 278; *New York Times,* December 27, 1935; Hawley, pp. 91–93; Lyon, pp. 711–12.

brother's keeper, and especially . . . the keeper of your small industrial brother."[22]

But the onslaught against fixed prices continued. In March the president responded to pressures to investigate NRA by appointing the National Recovery Review Board, headed by the famed criminal lawyer and champion of unpopular causes, Clarence Darrow. At the Darrow hearings, discontented industrialists and others testified that the NRA promoted monopoly and "oppressed small enterprises through price fixing. The NRA discounted as insubstantial some testimony against subdivisional coal code authorities in northern West Virginia and western Pennsylvania. Speaking as a member of the NRA's Labor Advisory Board, John L. Lewis blasted the Darrow board for having taken its data from "irresponsible malcontents, sweatshop employers and business interests which had lost special privileges." Yet the Darrow reports, covering thirty-four industries, came out in May and June, 1934, and they gave the NRA a black eye.[23]

That May, consumer advocates and antitrusters won a seeming victory when Johnson named Leverett S. Lyon of the Brookings Institution to head a new policy group. A previous critic of NRA pricing policy, Lyon determined to establish a national policy on code trade practices. His recommendations led on June 7, 1934, to Office Memorandum 228, which set down the official NRA price policy. The memorandum barred price fixing except in emergencies and situations of "destructive price cutting." In June, 1934, Borah and Nye had apparently caused the NRA to reverse itself and return to a free-market policy.[24]

22. Roosevelt, "Address before Code Authorities," March 5, 1934, *Public Papers*, vol. 3, pp. 123–30; Schlesinger, p. 132.

23. "NBCIB Hearing," January 3, 1935, p. 256; Lyon and Abramson, p. 28; U.S., Congress, Senate, Committee on Finance, *Investigation of the National Recovery Administration, Hearings Pursuant to S. Res. 79*, 1935, p. 117; Hugh S. Johnson, *The Blue Eagle from Egg to Earth* (New York: Doubleday, Doran and Co., 1935), p. 272; *New York Times*, May 19, 21, 1934; U.S., National Recovery Review Board, "First Report to the President of the United States" (mimeographed copy, Harvey Firestone Library, Princeton University, Princeton, N.J.), pp. 126–30; Roosevelt, vol. 2, pp. 136–37; Schlesinger, p. 132–34; Hawley, pp. 84–85, 96–97; Lyon, pp. 716–17.

24. Hawley, pp. 98–100; Miller, pp. 336–49; Schlesinger, p. 135; Lyon, chapter 26 and pp. 716–35; *Chicago Tribune*, editorial, May 28, 1934; Donald Richberg, *The Rainbow* (Garden City: Doubleday, Doran and Co., 1936), pp. 112–13.

Office Memorandum 228 threatened to undercut eight months of hard work to stabilize coal prices. Coal operators and other industrialists sent Deputy Administrator Ellis so many protests, a high NRA official joked, that he "couldn't get out of his office." Johnson claimed there had been a misunderstanding: the new policy did not apply to existing codes, but only to future ones.[25]

Establishing Market Areas

As consumer representatives won the policy struggle, coal industrialists evolved various practical schemes to make price fixing more effective. Following receipt of complaints that chiselers were breaking down coal prices in some markets, Johnson summoned the NBCIB into session in mid-January, 1934. Ill, Johnson sent in his place NRA General Counsel Donald Richberg. In language similar to that Roosevelt used to the representatives of the code authorities, Richberg warned the coal men that if they missed their opportunity to stabilize prices and degenerated "back into a disorderly, lawless, antisocial industry in which everybody tries to cut everybody else's throat," then further governmental action "can be expected." The point was clear: end interdivisional price disagreements or face a special law along the lines of the Lewis-Hayden bill.[26]

Although the operators were stabilizing prices in many markets, subdivisional code authorities and marketing agencies continued to compete among themselves. Some attempts to coordinate intersubdivisional prices collapsed. Unless interregional price conflicts were ended, *Coal Age* feared, the benefits of the code might disappear "under a wave of sectional jealousies." A plan to allocate tonnage among Appalachian operators using quotas of the average production of 1929–1933 failed. Ohio disliked the base years—times when they were one of the few regions paying

25. George Galloway, *Industrial Planning under the Code* (New York: Harper and Brothers, 1935), p. 61; official quoted in *Coal Age* 39 (July, 1934): 291; Lyon, pp. 735–38; Schlesinger, p. 135; Hawley, pp. 101–102.

26. Richberg statement, "NBCIB Hearing," January 16, 1934, pp. 1–3, 4, 14; *New York Times*, January 19, 1934; Hawley, p. 137.

union wages. Western Kentucky objected to its quotas. Western Pennsylvania withdrew.[27]

But despite all the radical twists in policy at the national level and interregional rivalry, coal operators finally began to develop a workable system for correlating prices among subdivisions. Instead of setting prices by mine groups and absorbing freight charges, they set them by market areas. The code authority for Ohio, for example, established eight market areas where Ohio coal traditionally sold. Prices of similar coal from different producers moving into any one market area might vary five or ten cents, reflecting mining conditions and freight rates. But market-area prices tended to minimize competition. By July 1, 1934, all the subdivisions of the Appalachian division had established market areas with firm prices.[28]

The beginning of the new coal year that April 1 further strengthened the profit position of the industry. The union won sizable wage increases and established the seven-hour day in the renegotiation of the Appalachian Agreement effective April 1. But the industry increased the percentage of contracts negotiated at code prices from about half to four-fifths. Average prices for the main eastern region went up twenty-two cents a ton and covered the wage increases.[29]

Just as operators stabilized coal prices, however, NRA leadership collapsed. The general's personal inclination to run a one-man show weakened morale among his deputies. His somewhat justified paranoia about "phantoms and nameless opposition" drove him to excesses of work and drink. Convinced that the NRA needed its own stabilization, Richberg, Tugwell, Ickes, and Perkins persuaded Roosevelt to get Johnson to take a long vacation in August, 1934. Newsmen wrote of the reorganization of the NRA under a board. In September, 1934, thinking that this publicity killed business co-

27. *Coal Age* 39 (July, 1934): 257, 291; (October, 1934): 400; (August, 1934): 325; "Minutes of Meeting of Representatives of Code Authorities of Divisions I, II, & III, held June 1, 1934, Netherlands-Plaza Hotel, Cincinnati, Ohio," CF, BC, Code Authority, Meetings, RG 9, NA; NRA, "Economic Survey," vol. 2, p. 524.

28. Ohio Code Authority Price Schedule No. 6 is reproduced in NRA, "Economic Survey," vol. 2, p. 515; see also vol. 2, pp. 512–16.

29. For the impact of wage changes, see chapter 6. NRA, "Economic Survey," chart 16, appendix 3.

operation and ruined public confidence, Johnson presented his own plans for reorganization of the NRA and resigned.[30]

In his place, Roosevelt appointed a five-man National Industrial Recovery Board (NIRB). Composed of two business leaders, Arthur Whiteside of Dun and Bradstreet and S. Clay Williams, former head of R. J. Reynolds Tobacco; Sidney Hillman for labor; and college professors Walton Hamilton and Leon C. Marshall, representing consumers and the public, the NIRB institutionalized the policy stalemate. Chaired by an able but staid Williams, the divided board tended to maintain the standoff between the antitrusters and those "cooperationists" who sought to maintain "price floors" and "self-government in industry." Younger NRA officials, however, pressed for change.[31]

Chief NRA economic and legal advisors Leon Henderson and Blackwell Smith, as ex officio members of the NIRB, took up the slack allowed by the NIRB stalemate. Thomas Blaisdell, who headed the Consumers' Division following Mary Rumsey's death in 1934, Henderson, and Smith argued for freeing prices to stimulate production. Many in the NRA believed that the code system had overextended itself. In his second fireside chat of 1934, Roosevelt noted that it was time to review whether the codes had "gone too far in such matters as price fixing and the limitation of production. . . ." Richberg in early September, 1934, counselled against "rash changes in price fixing policies." But he was shifting with the winds, and as director of a revised, policy-setting Industrial Emergency Committee he hinted at the end of price fixing in a National Press Club speech in early October.[32]

30. Richberg memorandum to Marvin McIntyre, August 18, 1934, Richberg Papers, Box 45; Richberg, *Rainbow*, pp. 74–75; Donald Richberg, *My Hero: The Indiscreet Memoirs of an Eventful but Unheroic Life* (New York: G. P. Putnam's Sons, 1954), pp. 166–76; Harold Ickes, *The Secret Diary of Harold Ickes*, 3 vols. (New York: Simon and Shuster, 1953–54), vol. 1, pp. 197–98; Johnson quoted in Hawley, pp. 104–106; Schlesinger, pp. 152–57; Sternsher, p. 167.

31. "NRA is Reorganized," in Roosevelt, vol. 2, pp. 405–407; Hawley, pp. 106–10; Schlesinger, p. 157.

32. "Industrial Emergency Committee is Created," June 30, 1934, in Roosevelt, vol. 2, p. 333; "Second Fireside Chat of 1934," in Roosevelt, vol. 2, pp. 417–21; Richberg to McIntyre, September 5, 1934, Box 45, Richberg Papers; Richberg, *Rainbow*, p. 170; Richberg, *My Hero*, pp. 183, 194; Hawley, p. 107; Schlesinger, pp. 158–62; Thomas Vadney, *The Wayward Liberal* (Lexington: The University of Kentucky Press, 1970), chapter 8.

The appointment of the NIRB and the public statements by Richberg and Roosevelt implying the end of price and production controls symbolized the NRA's general retrenchment. The constitutionality of the coal operators' price-fixing program had been upheld in the courts. But the NRA's seeming move toward the free-market system weakened the legality of the coal industry's minimum prices. As the government publicly admitted its own uncertainty, coal purchasers in Pittsburgh, Buffalo, and Charleston, West Virginia, stopped negotiating contracts in the hope that minimum coal prices would be abolished. Coal prices began to break. Throughout the country operators began negotiating contracts for post-code delivery at less than the code levels. The western Pennsylvania code authority officially resolved to meet any competitive subdivision's prices. Other code authorities protested to the NIRB that changing national policy would destroy the gains of the code.[33]

The NIRB told the coal men not to worry. One code authority secretary wrote Wayne P. Ellis, now divisional administrator for coal: "The code authorities had a right to feel greatly alarmed. The bituminous coal industry is entirely surrounded by a bunch of buzzards consisting of jobbers, wholesalers, the buying public and chiselers who promptly seize upon anything that is published which has a tendency to in any manner break down either fair minimum market prices or code compliances. . . ." Code authority leaders formally requested the NIRB to state that the "Bituminous Coal Code of Fair Competition and all of its provisions are in full force and effect. . . ."[34]

John L. Lewis moved vigorously to shore up the crumbling price levels. He estimated that up to 50 percent of the operators were signing contracts at lower than code prices for coal to be delivered after the legal expiration of NRA in June, 1935. Newspapers reported that Massachusetts public bids

33. NRA, "Economic Survey," vol. 2, p. 522; *New York Times,* October 6, 1934; *Coal Age* 39 (September, 1934): 365; (December, 1934): 505.

34. See the file of telegrams to Ellis in CF, BC, Prices, Miscellaneous, RG9, NA; David C. Rea, secretary, Northern West Virginia Code Authority, October 8, 16, 1934, CF, BC, Code Authority, Meetings, RG9, NA: "Resolution, October 8, 1934," CF, BC, Division I, Code Authority, RG 9, NA.

were being made at from thirty to fifty cents below code prices. On December 17, 1934, aware that low prices would pressure wages, Lewis told Division Administrator Ellis that the NRA's failure to "meet this menacing situation" would compel the UMW "to take such steps as may be deemed necessary to protect the interests of its members." The operators, Lewis wrote, were following a "policy of monumental stupidity" and seemed "incapable of preventing their own commercial destruction." He suggested that NRA once again assemble the NBCIB to discuss the difficulties.[35]

The NBCIB met in early January, 1935, concurrent with the NRA's second general price hearing, and voted to make post-code contracts below the current fair-market prices a violation of the code and to allow the UMW to sit on code authorities.[36] But how to shore up the price-fixing apparatus? Some wanted more government control over business, and others wanted industrial "self-government." Blackwell Smith's lawyers drafted a coal code amendment empowering the NIRB "through such agencies as it may designate," to investigate costs and "determine and publish stated minimum prices." The National Coal Association's legislative committee preferred leaving price fixing in operator hands but wanted to create a structure for arbitrating interdivisional price disputes.[37]

In the NBCIB discussions, one NRA spokesman hinted darkly that if the operators refused governmental price fixing and cost investigation, NRA would remove all price stabilization protection. Lewis retorted that other industries had "preyed upon" the coal industry and "sucked its heart's blood." They will, he said, "immediately emerge from their hiding places and raid the defenseless industry again." Perhaps, as *Business Week* surmised, this government threat may have been just a ploy to get the operators to resolve

35. Lewis to Ellis, December 17, 1934, CF, BC, Prices, RG9, NA, reprinted in "NBCIB Hearings," January 4, 1935; *New York Times*, December 21, 1934; Charles O'Neill testimony, House Committee on Ways and Means, *Stabilization*, p. 160.

36. "NBCIB Hearing," January 4, 1935, pp. 71–72, 387–88; January 11, 1934, pp. 35–36; NRA, "Economic Survey,' vol. 1, pp. 125–31.

37. NRA, "Economic Survey," vol. 1, pp. 125–26; *New York Times*, December 27, 1934; Hawley, pp. 107–108.

their own price-coordination problems. But the operators took the threat seriously. If they did not pass the NCA proposal, they thought they faced special legislation and government price fixing or a possible strike by the union.[38]

The NBCIB operators first outvoted the presidential members and defeated the government's proposal. Then, after securing the approval of the NRA legal, industrial, labor, and consumers' advisory boards, they unanimously passed the NCA amendment. Boards in each code division would now arbitrate interregional price disputes. In establishing prices, code authorities must consider a new cluster of variables: the purposes of the NRA, wage rates, the protection of investors, other energy competition, trade customs, consumer buying habits, and chemical and physical properties of the coals. Prices so established were to be published, with hearings for complaints. Appeal from a code-authority decision would run to the local arbitration boards and then to an impartial National Coal Board of Arbitration.[39]

The National Coal Board of Arbitration came too late. Arbitration might have succeeded in the fall of 1933 if the operators had been willing then to create such a board, but they were not. In the spring of 1935, with the legal end of the code fast approaching, the board had little chance to work. Established quickly in the Shoreham Hotel, the board promulgated a set of rules to govern cases, most specifically noting that it would consider only matters whch had been discussed at a lower level and submitted in writing. Most of the forty-five controversies handled through the board during its short history came from the Appalachian territory. Many agreed with Charles O'Neill when he noted that the board was "ineffective."[40]

38. "NBCIB Hearing," January 3, 1935, p. 116; January 9, 1935, pp. 8–12; January 3, 1935, pp. 113, 273; January 11, 1935, p. 20; and January 16, 1935, pp. 1–10. *Business Week*, January 12, 1935, p. 20; *New York Times*, January 5, 1935; NRA, "Economic Survey," vol. 1, p. 127.

39. "NBCIB Hearings," January 10, 1935, passim; January 11, 1935, pp. 18, 29; NRA, "Economic Survey," vol. 1, p. 128; vol. 2, pp. 527–28; *Coal Age* 40 (February, 1935): 91–93.

40. "National Coal Board of Arbitration, Rules and Regulations," Feburary 9, 1935, in Cf, BC, Code Authority, N.C.B. A, RG9, NA; William Jett Lauck Diary, May 8, 1935, Lauck Papers: "Bituminous Coal Industry General Conference, April 4, 1935," pp. 4, 25, Library Unit Records, Transcripts of Hearings, Rg 9, NA.

As more and more operators contracted for post-code sales below the minimums, Ellis called for a conference in April, 1935, to "stiffen up the price situation." Operator statements at the conference reflected the dismay of those who had hoped to make NRA prices effective. Having taken over Ellis's job as deputy administrator for coal, Newell Roberts admitted that the code had "nearly broken down." The Supreme Court's decision in the case of *Schechter* v. *United States,* which invalidated the National Industrial Recovery Act on May 27, 1935, administered the *coup de grâce* to the attempts to establish minimum prices for the soft-coal industry. The industry was left after *Schechter* where it had begun back in 1933, faced with the prospect of price wars.[41]

Conclusions

Despite the breakdown of the code prices during the last months of the NRA, the operators' success in boosting prices some 37 percent under the code remains remarkable, given the difficulties they faced (see Table 4). The system of determining prices by market areas did not eliminate the competition among regions, but it provided enough monopoly control to be effective in the short run. The increased labor costs of the Appalachian agreements of 1933 and 1934 provided a cost base that encouraged operators to market their coals at higher prices. NRA figures showed margins between average price and average costs (excluding capital charges) to be only 2.6 cents per ton.[42]

The qualified success of the bituminous coal code in stabilizing prices became a matter of great pride for the NRA. An October, 1934, nationwide poll of operators showed overwhelming support for price controls. Johnson bragged

41. "Bituminous Coal Industry General Conference, April 4, 1935," pp. 42–43, Library Unit Records, Transcripts of Hearings, RG 9, NA; House Committee on Ways and Means, *Stabilization,* p. 160; Senate Committee on Interstate Commerce, *Stabilization,* p. 73; William Lauck, "Docket—United Mine Workers," March 8, 1935, Lauck Papers; *Schechter* v. *United States,* 295 U.S. 495 (1935).

42. NRA, "Economic Survey," vol. 2, p. 550; Parker, p. 127; Richberg testimony, Senate Committee on Finance, *Investigation of NRA,* pp. 83, 111; Van Bittner testimony, Senate Committee on Interstate Commerce, *Stabilization,* p. 137; Newcomb, p. 24.

Table 4. Bituminous Prices and Production, 1929–1935

Year	Average Per-Ton f.o.b. Mine Price	Production (tons)
1929	$1.78	534,989
1930	1.70	467,526
1931	1.54	382,089
1932	1.31	309,710
1933	1.34	333,631
Jan.–Oct.	1.29*	
Nov.	1.56	
Dec.	1.59	
1934	1.75	359,368
Jan.	1.63	
Feb.	1.64	
Mar.	1.63	
April–Nov.	1.86	
Dec.	1.91	
1935	1.77	372,373
1936	1.76	439,088
1937	1.94	445,531
1938	1.95	348,545
1939	1.84	394,855
1940	1.91	460,772
1941	2.19	514,149

SOURCES: U.S., Bureau of the Census, *Historical Statistics of the United States: Colonial Times to 1957* (Washington, D.C.: Government Printing Office, 1960), p. 356, and U.S., National Recovery Administration, "Economic Survey of the Bituminous Coal Industry under Free Competition and Code Regulation" (mimeographed), by F. E. Berquist, et al., Work Materials No. 69, 2 vols. (Washington, D.C., 1936), chart 16, app. 3. The monthly figures are for the three major geographic divisions under the code structure. NRA did not collect accurate data during the last four months of its existence.
*estimated

that the coal code was "the finest product" of the NRA. Richberg told congressional investigators in early 1935, "I think that the record of the bituminous coal industry in the code, and under all of its difficulties, it has been a very difficult and embarrassing code, is an outstanding and major achievement. . . ." He thought that 90 percent of the operators favored maintaining the code's protection.[43]

43. *Coal Age* 39 (October, 1934): 380–82; Richberg testimony, Senate Committee on Finance, *Investigation of NRA*, pp. 112–13; Johnson quoted in "The Coordinator" (mimeographed newsletter of the Smokeless Code Authority), July 26, 1934, p. 2, CF, BC, Smokeless Code Authority, Organization, RG 9, NA.

One leading southern operator who had opposed the NIRA originally noted that "after it was enacted and the industry was compelled to operate under it, the results were so surprisingly favorable that I reluctantly became a convert to governmental control." The *Mining Congress Journal* noted that "even the most ardent exponents of 'rugged individualism' admit that the code has been beneficial and that the industry is emerging from at least some of its chaos." James D. A. Morrow, whose Pittsburgh Coal Company showed the best earnings since 1924, proclaimed, "You can call it government control, state socialism, or whatever you like. I don't care. It's the practical answer for those who have money invested in the business and for those who are employed in the industry." The NRA Committee on Industrial Analysis reported that of the codes that attempted minimum price fixing, bituminous coal had achieved the most success.[44]

As was the case with NRA wage-and-hour gains, however, the short-run accomplishments of the industry in raising prices had a deleterious long-range impact. Higher prices for coal encouraged consumers to shift to oil, water power, and natural gas, although high switching costs and the minor part coal costs played in total railroad or utility costs delayed this shift until after the war.

The NRA unfortunately had gone about reviving the economy the wrong way. Indeed, the "success" in raising coal prices increased industrial production costs and probably hindered general economic recovery. As in the case of wages, price stabilization in the coal industry worked as a holding action, providing a respite from the ravages of the Great Depression but not really solving anything. Marginal operations were kept in business, and more economic reallocation of resources was postponed. As Table 3 and Fig. 1 show (pp. 192 and 20), only the upsurge of war-stimulated demand would achieve the real goal of increased productivity in coal mining.[45]

44. E. C. Mahan testimony, U.S., Congress, Senate, Committee on Interstate Commerce, *Hearings on S. 4668, A Bill to Regulate Interstate Commerce in Bituminous Coal*, 74th Cong., 2d sess., 1936, p. 233; *Mining Congress Journal* 21 (March, 1935): 14; Morrow quoted in *New York Times, August 26, 1934;* NRA, *Committee of Industrial Analysis*, "Report," pp. 166–68.

45. Lyon et al., p. 873.

The industry's profits did grow under NRA minimum prices. Although there are no reliable industry-wide figures on return-on-investment or general profits, aggregate data from the Internal Revenue Service give an outline of the industry's status during the depression. The industry reported a net deficit every year from 1928 to 1936. A $42 million deficit in 1930 grew to $47.7 million in 1931 and to $51.1 million in 1932. It then shrank to $47.5 million in 1933. In the first full year under the NRA the deficit fell to $7.6 million. New mines opened, production increased 7.7 percent, and wages and prices both increased. The net deficit grew in 1935 to $15.5 million and then declined in 1936 to $3.3 million. Not until 1940 would the industry report a net profit.[46]

Stabilizing bituminous coal prices had been one of the NRA's most ambitious undertakings. Given the industry's competitive structure and regional nature, it seems miraculous that the code authorities and marketing agencies were able to stabilize prices at all. As will be seen in the chapter on the Guffey acts, by comparison the NRA accomplished much. Industrial self-government had worked for a period of months, only to be destroyed by regional competition. Those who wanted to continue some kind of governmental intervention to stabilize the industry, however, would soon have other proposals to offer.

46. Ellery B. Gordon and William Y. Webb, *Economic Standards of Government Price Control*, TNEC Monograph 32, (Washington, D.C.: Government Printing Office, 1940), p. 334; Waldo E. Fisher and Charles M. James, *Minimum Price Fixing in the Bituminous Coal Industry* (Princeton: Princeton University Press, 1955), p. 18.

8

The Failure of the Guffey Acts

DURING THE SPRING of 1934, John L. Lewis and Pittsburgh Coal's James D. A. Morrow renewed their efforts to get a coal stabilization bill through Congress. The National Industrial Recovery Act, which had supplanted earlier special coal bills, could be extended past its expiration date of June 16, 1935. But the two men wanted a bill which offered more centralized controls than NIRA had provided. Lewis dispatched his chief legislative draftsman, Henry Warrum, on a European trip to study British and continental mine regulation. Morrow began the difficult job of herding operators from various regions toward a common bill which the National Coal Association committee on legislation could support.[1]

The UMW representatives met with the NCA committee during the summer and early fall to see if labor and management could agree on a bill. They all wanted a new regulatory commission with power to stabilize prices. They also began to consider a proposal to retire excess high-cost mine properties then being developed by Roosevelt's Planning Committee for Mineral Policy.[2]

1. J. D. A. Morrow to John L. Lewis, September 8, 1934; UMW Vice President Philip Murray, memorandum, September 29, 1934, both in box for 1934–36: Stabilization Bill, Brophy Papers. Ellis W. Hawley, *The New Deal and the Problem of Monopoly* (Princeton: Princeton University Press, 1966), p. 189.

2. *New York Times*, January 13, 1934; Arthur M. Schlesinger, Jr., *The Age of Roosevelt*, vol. 2, *The Coming of the New Deal* (Boston: Houghton Mifflin, 1958), pp. 545–48; Glen Lawhon Parker, *The Coal Industry: A Study in Social Control* (Washington, D.C.: American Council on Public Affairs, 1940), pp. 136–37.

But the age-old regional differences among operators— primarily the sharp North-South cleavage which separated the former union territories of the Midwest from the South— and the operator-union split prevented agreement. In testimony to the operators' new-found support for government regulation, the October, 1934, NCA annual convention approved the Morrow committee's general recommendation to seek "special legislation" for the industry. But lacking a definite legislative proposal, the delegates endorsed a two-year extension of NIRA for the moment. The NCA endorsed the idea of price and wage controls and declared that "the bituminous coal code has been a forward step toward stabilization and rehabilitation of the industry...."[3]

Business Week praised the coal men—the first industrialists to begin planning beyond the expiration date of NIRA. John L. Lewis asked operators for their further suggestions for special legislation and said that the union was determined to seek congressional action on a coal bill "designed to help the coal operators just as much as it is designed to help the mine workers." In early 1935, using an introduction from a western Pennsylvania operator, Lewis and Henry Warrum presented the union's bill to Pennsylvania's new freshman senator, Joseph F. Guffey.[4]

Incorrectly called a "little NRA," the Guffey bill carried out the general objectives of NRA but drew on entirely new sources and went far beyond the code. A National Bituminous Coal Commission of five impartial members would promulgate a code of fair competition. In a section

3. *New York Times*, October 28, 1934; *Business Week*, November 3, 1934; *Coal Age* 39 (November, 1934): 415–16.

4. *Business Week*, November 3, 1934, p. 14; "Hearing before the National Bituminous Coal Industrial Board," (henceforth cited as "NBCIB Hearing"), January 4, 1934, pp. 89–90; January 9, 1934, pp. 13–15; January 10, 1934, p. 17 (typescripts), Library Unit Records, Transcripts of Hearings, RG9, NA; operator Howard Essington and John Lewis testimony, U.S., Congress, House, Committee on Ways and Means, *Stabilization of the Bituminous Coal Mining Industry, Hearings*, 74th Cong., 1st sess., 1935, pp. 219, 648; Guffey testimony, U.S., Congress, Senate, Committee on Interstate Commerce, *Stabilization of the Bituminous Coal Mining Industry, Hearings Before a Subcommittee*, 74th Cong., 1st sess., 1935, p. 168; William Jett Lauck, "Docket—United Mine Workers," January 9, 17, Lauck Papers; Joseph P. Guffey, *Seventy Years on the Red-Fire Wagon* (Lebanon, Pa.: published by the author, 1952), pp. 43–46, 74–77, 82–83, 132; Ralph Hillis Baker, *The National Bituminous Coal Commission* (Baltimore: Johns Hopkins University Press, 1949), pp. 48–49.

that stemmed from Warrum's study of the British Coal Mines Act of 1930, a National Coal Producers Board would fix minimum prices and allocate tonnage among districts in the hope of resolving the regional conflicts which had weakened the code's price fixing. To encourage participation, 99 percent of an industry-wide tax of 25 percent on all coal would be rebated only to those agreeing to comply. The bill included Section 7a of the NIRA and stated that any wage contract negotiated with operators producing two-thirds of the tonnage in a district or group of districts would be binding on all code members in that area. Allocation of tonnage, centrally determined prices, severe tax penalties, and a kind of regional "union shop" made the Guffey bill a much tougher, more centralized plan than the code.[5]

Clearly a special-interest proposal, this union–northern operator scheme seemed the only port in a storm to many in the coal industry. But like similar examples of group-sponsored legislation in American history, however liberal it may have seemed at the time, the Guffey coal bill was economically short-sighted and basically conservative. Attempting to revive a depressed sector of the economy by centrally regulating prices and allocating production would merely freeze the industrial status quo. It would perhaps save the northern operators from further inroads by the South, but it would hardly spur development of the industry. The Guffey bill was good politics for Lewis and the northern operators—and perhaps for the New Deal. Unfortunately, it was economically unsound and—after repassage as the Guffey-Vinson Act—became a bureaucratic boondoggle.

The enactment of two Guffey acts testified to the strength of the revived UMW, to the union's clout with the New Deal, and to the alliance between John L. Lewis and the operators in the Central Competitive Field. In combination with the union, the coal operators of the North finally had major influence on government. Allied with Lewis, they secured two special laws to stabilize—and unionize—the industry. Neither, however, worked.

5. Brophy, Oral History Memoir, p. 541, Brophy Papers; *Coal Age* 41 (February, 1936): 44; *New York Times*, January 26, 1935; Guffey, pp. 92–93.

The First Guffey Act

At the spring, 1935, hearings, Central Field operators—representing the same areas that had sent representatives to President Wilson—backed the bill. Charles O'Neill, the former coal association leader from Pennsylvania, enlisted Pennsylvanians, some Ohioans, many from Illinois, a few from Indiana, and scattered producers elsewhere who wanted to stabilize the entire industry and/or halt the growth of the southern wing of the industry. Union wages and allocation of tonnage based on 1919–1934 averages clearly favored the Central Field. Southerners, the Rockefeller and Morgan coal interests, some westerners, and other operators not ready to capitulate to the power of the UMW–Central field alliance, opposed the bill. West Virginian James D. Francis, one of the "four horsemen" of the industry who had hammered out the original code, called the bill a straitjacket. Alabaman Darius Thomas predicted the "complete annihilation of the coal industry."[6]

A host of conservatives, increasingly outraged by the New Deal, joined the anti-Guffey army. The National Association of Manufacturers, the U.S. Chamber of Commerce, owners of the captive mines, the railroads, and other large consumers all attacked the measure. While finding much that was appealing, *Coal Age* urged that Congress extend the NIRA, which had "effected such a beneficial revolution in the bituminous coal industry," rather than experiment with "an importation from Great Britain."[7]

6. *New York Post,* June 12, 1935; *Washington News,* May 21, 1935; *Baltimore Sun,* May 29, 1935; *Coal Age* 40 (April, 1935): 169; (August, 1935): 354–56; Parker, p. 169; testimony of operators Forney Johnston, Henry DeBardeleben, James D. Francis, H. L. Findlay, Jonas Waffle, and Charles O'Neill, House Committee on Ways and Means, *Stabilization,* 164, 446, 460–61, 495–96, 519, 603–604; testimony of operators H. R. Hawthorne, James W. Carter, James D. Francis, P. C. Thomas, Kenneth A. Spencer, E. H. Suender, in Senate Committee on Interstate Commerce, *Stabilization,* pp. 181–90, 220–36, 313–15, 393, 442–44, 504; J. V. Sullivan, "Coal Producers Oppose Guffey Bill," *Manufacturer's Record* 104 (May, 1935): 32.

7. W. Jett Lauck, "Coal Labor Legislation: A Case," *Annals* 184 (March, 1936): 132–36; John L. Lewis testimony, House Committee on Ways and Means, *Stabilization,* pp. 633–34; Philip Murray testimony, Senate Committee on Interstate Commerce, *Stabilization,* p. 153; *Baltimore Sun,* May 11, 1935; *New York Times,* May 26, 1935; *New York Journal of Commerce,* June 17, 1935; James T. Patterson, *Congressional Conservatism and the New Deal: The Growth of a Conservative*

Abandoning the principles of NRA, the president stated on February 20, would be "unthinkable." In an obvious reference to coal, Roosevelt proclaimed that "certain natural resources" needed government supervision to eliminate waste, control output, stabilize employment, and end ruinous price cutting. On May 25 the White House indicated that Roosevelt would press for the coal bill, but only after the Congress extended the NIRA for the rest of industry.[8]

The Supreme Court decision in *Schechter* v. *United States* of May 27 rendered all talk of extending the NIRA moot. In a unanimous decision written by Chief Justice Charles Evans Hughes, the court declared that Congress had exceeded its power in delegating authority to the Executive Branch and to business code-drafting committees without providing the necessary standards for them to follow; moreover, the Schechters' poultry business was intrastate and had no direct effect on interstate commerce. The *Schechter* decision forced coal operators and legislators to scrutinize carefully the constitutional aspects of the Guffey bill.[9]

The day after the *Schechter* decision, the ongoing negotiating sessions for the 1935 Appalachian wage contract adjourned without agreement. The talks had begun in February with Lewis demanding the thirty-hour week and a 10 percent wage increase. Operators rejected the UMW proposal, which they said would increase costs some forty-five cents a ton. When negotiations snagged in late March, the NRA got both sides to extend the 1934 agreement beyond the April 1, 1935, expiration date. In accepting the prolonga-

Coalition in Congress, 1933–1939 (Lexington: University of Kentucky Press, 1967), p. 60; William E. Leuchtenburg, *Franklin D. Roosevelt and the New Deal, 1932–1940* (New York: Harper and Row, 1963), pp. 147–62; *Coal Age* 40 (March, 1935).

8. Lauck, "Docket—United Mine Workers," April 2, May 1, 25, 1935; *Coal Age* 40 (March, 1935): 132; *Vital Speeches* (March 11, 1935): 361; Franklin D. Roosevelt, *Complete Presidential Press Conferences of Franklin D. Roosevelt,* 25 vols., February 13, 1935, vol. 5, pp. 105–106; May 1, 1935, vol. 5, pp. 257–58.

9. A. L. A. *Schechter Poultry Corp. et al.* v. *United States,* 295 U.S. 553; Donald G. Morgan, *Congress and the Constitution: A Study of Responsibility* (Cambridge, Mass.: Harvard University Press, 1966), p. 170; Merlo J. Pusey, *Charles Evans Hughes,* 2 vols. (New York: The Macmillan Co., 1951), vol. 2, pp. 738–41. *Coal Age* 40 (June, 1935): 233–34, 271; *Washington Times,* May 28, 1935; Lauck, "Docket— United Mine Workers," May 27, 1935.

tion to June 16, Lewis hinted that the UMW would not accept a contract until Congress either extended the NIRA or passed the Guffey bill. By breaking off negotiations the day after the *Schechter* decision, the union operators and the UMW began a unified campaign to pressure Roosevelt and Congress to pass the Guffey bill.[10]

On June 14, two days before the strike which would come when the extended wage agreement expired, Roosevelt told the spokesmen for the two sides in the wage negotiation that he would support the Guffey bill in exchange for a further two-week extension. Bereft of the NIRA, under attack from conservatives on the right, and pressed to act by swarms of Democrats elected in the 1934 election, Roosevelt went on record for the Wagner labor bill, a new banking bill, the public utility holding company bill, the social security bill, and a wealth tax proposal. Making a potential model for other natural resource industries, Roosevelt called the Guffey bill "no. 1 must."[11]

Aggressive southern operator lobbying and serious doubts about its constitutionality, however, kept the bill in subcommittee in the House of Representatives. When the twice-extended wage contract deadline approached, White House emissaries persuaded Lewis to accept a third extension— this time of thirty days, to the end of July. Then J. D. A. Morrow and other former supporters rejected the bill on

10. *Coal Age* 40 (June, 1935): 273; U.S., NRA, "Economic Survey of the Bituminous Coal Industry under Free Competition and Code Regulation" (mimeographed), by F. E. Berquist et al., NRA Work Materials No. 69, 2 vols., Washington, D.C., vol. 2, pp. 327–28; *Washington Herald*, May 29, 1935; *New York Times*, February 18, 1935; Lauck, "Docket—United Mine Workers," February 20, March 13, 28, April 2, 11, June 20, 1935. Joseph Guffey to Marvin McIntyre, March 21, 1935; Donald Richberg to Marvin McIntyre, April 16, 1935; Committee Against the Guffey Bill to Virginia Ellis Jenckes, June 19, 1935, all in Official File (henceforth cited as OF) 175, Box 3, Roosevelt Papers; John Brophy, Oral History Memoir, chapter 17.

11. James MacGregor Burns, *Roosevelt: The Lion and the Fox* (New York: Harcourt, Brace & World, 1956), pp. 219, 224–26; Leuchtenburg, chapter 7; Arthur M. Schlesinger, Jr., *The Age of Roosevelt*, vol. 3, *The Politics of Upheaval* (Boston: Houghton Mifflin, 1963), chapter 21; *Vital Speeches* 1 (June 17, 1935): 492; *New York Times*, June 2, 3, 6, 14, 17, 1935. Marvin McIntyre, memorandum for Mr. Charles West, Farm Credit Administration, June 16, 1935; Roosevelt to Harold Ickes, June 17, 1935, both in OF 175, Box 3, Roosevelt Papers; *Newsweek*, June 22, 1935, p. 9; Lauck, "Docket—United Mine Workers," June 14, 1935; Roosevelt, *Presidential Press Conferences*, June 28, 1935, vol. 5, p. 393.

constitutional grounds. Attorney General Homer A. Cummings admitted that he had reservations about the bill's constitutionality. Roosevelt then wrote subcommittee Chairman Samuel B. Hill not to "permit doubts as to constitutionality, however reasonable, to block the suggestion."[12]

Roosevelt intervened in the wage talks for a fourth time and won an extension to September 16. The subcommittee stripped the coal reserve section from the bill and reported the bill without recommending its passage.[13] Working closely with Ways and Means Committee Chairman Robert Doughton of North Carolina, Roosevelt held wavering Democrats in line long enough for the committee to report the bill to the House by a one-vote margin—again without recommending passage. Six days later, as Congress hastened to adjourn, the bill slipped through, 194–168. Nearly one hundred Democrats deserted the president. When the measure cleared the Senate, 45–37, three days later, there were serious defections from Democratic ranks.[14]

Except for the labor provisions, the bill, which Roosevelt signed on August 30, 1935, was gutted. Allocation of production, the producers' board, and a coal reserve idea were scrapped entirely; districts rather than a central agency would determine prices. Consumers introduced a consumers' counsel; the rebate was lowered to 90 percent.[15]

The Guffey-Snyder Act of 1935 passed only because a strike threat orchestrated by Lewis and his operator supporters in the old Central Competitive Field got Roosevelt

12. *New York Times*, June 25, 30, July 1, 2, 6, 1935; *Coal Age* 40 (August, 1935): 353; Parker, p. 151; Morrow testimony, House Committee on Ways and Means, *Stabilization*, pp. 268–69; Morgan, pp. 182–83; Roosevelt to Samuel B. Hill, July 6, 1935, copy, OF 175, Box 3, Roosevelt Papers.

13. Lauck, "Docket—United Mine Workers," July 23, 1935, August 1, 9, 12, 16, 17, 18, 1935; *New York Times*, July 27, 31, 1935.

14. *Business Week*, July 6, 1935, p. 21. The House vote was four to one against in New England, four to three against in the normally Democratic South, six to five for in the Midwest, five to three for in the Mid-Atlantic region, and two to one for in the Border States and Far West. Kentucky split evenly. Patterson, pp. 60, 71; Morgan, pp. 173, 428; *New York Times*, August 16, 17, 1935.

15. Parker, p. 137; *Coal Age* 40 (July, 1935): 306–307; O'Neill testimony, House Committee on Ways and Means, *Stabilization*, pp. 2, 73–74, 153–57, 163–64; *Business Week*, July 6, 1935, p. 21; *New York Times*, June 9, 11, 1935; Waldo Fisher and Charles M. James, *Minimum Price Fixing in the Bituminous Coal Industry* (Princeton: Princeton University Press, 1955), pp. 30–31.

to pressure the liberal Seventh-Fourth Congress. Similar proposals for "Guffeyizing" other highly competitive industries like lumber and apparel lacked group backing and never became law. Despite the Guffey Act's emasculation, however, Lewis claimed that act was a "bill of rights" for the industry. New York Congressman Bertrand Snell declared it was "the first time in the history of the country that the House passed a bill which was opposed by the subcommittee which wrote it, by the full committee which approved it, by the Rules committee which brought it out on the floor, and toward which the entire house organization was entirely indifferent." Said Democratic Senator Millard Tydings, "Like an autumn flower it will be blown away by the first winter blast of the court."[16]

Carter v. Carter Coal Company

Within hours after Tydings' remark, fellow southerner James Carter sought to get the District of Columbia Supreme Court to enjoin the Carter Coal Company, of which he was the president and chief stockholder, from paying the coal tax. Filed by Frederick H. Wood—the successful defender of the Schechter brothers—the Carter brief asserted that Carter's business was intrastate, that the tax was a penalty, that the act invaded the rights reserved to the states or to the people by the Tenth Amendment and the Fifth Amendment's protections of property without due process, and that the act was "wholly arbitrary, capricious, and unequal."[17]

Although declared premature by the court on the grounds that the tax was not due until November, Carter's suit and approximately eighty other similar cases sabotaged the Guffey Act before it could begin.[18] Roosevelt made some politically determined, lackluster appointments to the National

16. Hawley, pp. 224, 279; *New York Times*, August 20, 21, 1935; *Washington Post*, January 25, 1935; *United Mine Workers Journal*, September 15, 1935, p. 14.

17. *Charleston* (West Virginia) *Gazette*, August 31, 1936; *New York Times*, September 1, 1935; *Time*, March 23, 1936, p. 23; Stanley Reed memorandum to Marvin McIntyre, August 31, 1935, OF 1732, Box 1, Roosevelt Papers; *Coal Age* 40 (October, 1935): 249.

18. *Time*, November 25, 1935, p. 19; *New York Times*, September 11, 1935; *Coal Age* 40 (October, 1935): 420.

Bituminous Coal Commission (NBCC), and the agency began its job of seeking operators to register under the act. But many of the larger firms, northern and southern, refused to sign. Following a filibuster of the appropriation for the agency, the NBCC borrowed furniture from the Interior Department, but its general orders to districts to develop cost statistics on which to base minimum prices brought little result. Operators held back and awaited the legal validation of the act. Of the ten million dollars assessed operators under the act to May 1, 1936, the NBCC received only eight hundred thousand dollars.[19]

Following some ambiguous lower court decisions, the Supreme Court agreed to take the case on certiorari at the beginning of the January, 1937, term. A negative decision seemed inescapable. The Court had invalidated the Agricultural Adjustment Act on the ground that the tax was a part of a regulatory system not for the general welfare and was a violation of the Tenth Amendment. A restrictive interpretation of the commerce clause had voided the NIRA and the Railroad Pension Act, and a restrictive definition of the power of delegation had wiped out the "hot oil" provision of the NIRA. The Supreme Court had accepted a stockholder's suit similar in form to Carter's in the TVA case.[20]

Because the Guffey Act rested on both the taxing and the commerce powers and raised questions of interstate commerce of far greater dimension than those in Brooklyn's poultry trade—including the validity of any federal regulation of wages and hours—enormous anticipation surrounded

19. *New York Times*, September 21, 1935, January 25, 1936; *Coal Age* 40 (November, 1935): 465; (December, 1935): 551–52; Lauck, "Docket—United Mine Workers," September 19, 20, 23, 1935; Joseph Guffey to James A. Farley, September 6, 1935, OF 1732, Roosevelt Papers; National Bituminous Coal Commission (henceforth cited as NBCC) General Orders Nos. 1—5, RG 150, NA: U.S., Congress, Senate, Interstate Commerce Committee, *Hearings on S. 4668: A Bill to Regulate Interstate Commerce in Bituminous Coal, and for Other Purposes*, 74th Cong., 2d sess., 1936, p. 18. Reports on Code Acceptances, RG 150, NA; R. C. Shriver to Charles E. Smith, December 18, 1935, Box 20, Smith Papers; Parker, pp. 140–41; Baker, pp. 58–59.

20. *New York Times*, November 1, 12, December 21, 24, 1935; January 9, 1936; *U.S.* v. *Butler*, 297 U.S. 1 (1936); *Schechter* v. *U.S.*, 295 U.S. 495 (1935); *Retirement Board* v. *Alton R. R. Co.*, 295 U.S. 330 (1933); *Panama Refining Co.* v. *Ryan*, 293 U.S. 338 (1935); *Ashwander* v. *T.V.A.*, 297 U.S. 288 (1936).

the Court's decision in *Carter* v. *Carter Coal*.[21] Speaking for the majority of five, Justice George Sutherland declared that the Tenth Amendment prohibited the federal government from assuming regulatory powers not enumerated in the Constitution. Shouldering aside the "stream of commerce" doctrine elucidated by Justice Oliver W. Holmes in *Swift and Company* v. *U.S.*, Sutherland cited *U.S.* v. *E. C. Knight*, the 1890's "Sugar Trust" decision, which held that "commerce succeeds to manufacture, and is not part of it. . . ." So it was with coal mining, said Sutherland. "The employment of men, the fixing of their wages, hours of labor and working conditions, the bargaining in respect of these things . . . is purely local" and unreachable by the federal government under the commerce clause.[22]

Having "reduced the 'stream of commerce' to a trickle," Sutherland disregarded the congressional statement in the act's preamble that the price sections were separable from the wage sections. Although he refused to consider the constitutionality of the price sections per se, he used the government's argument that the wages and prices interacted upon each other to strike down the law. Congress, he said, deemed both "necessary to achieve the end sought." Thus, the price provisions "are so related to and dependent upon the labor provisions as considerations or compensations, as to make it clearly *probable* that the latter being held bad, the former would not have passed" [italics mine].[23]

In a concurring opinion, Chief Justice Charles Evans Hughes objected to Sutherland's refusal to accept the statement of Congress that the wage and price sections were separable. "Trying to imagine what Congress would have done" if part of a statute were declared invalid, he chided, "leads into the realm of pure speculation." But he concurred

21. *Time*, March 23, 1936, p. 23; *New York Times*, March 5, 8, 10, 11, 13, 1936; *James Walter Carter* v. *Carter Coal Company* (transcript of proceedings), Miscellaneous Records, RG 150, NA, p. 102; *Literary Digest* 121 (May 30, 1935): 5.

22. *U.S.* v. *E. C. Knight*, 156 U.S. 1 (1895); *Carter* v. *Carter Coal Company*, 298 U.S. 242 (1936). Forrest Revere Black, "The Commerce Clause and the New Deal," *Cornell Law Quarterly* 20: 179.

23. *Nation* 142 (June 3, 1936): 700; Alpheus T. Mason, *Harlan Fiske Stone: Pillar of the Law* (New York: The Viking Press, 1956), p. 419; *Carter* v. *Carter Coal Company*, 298 U.S. 316.

"that production—in this case mining—which precedes commerce, is not itself commerce; and that the power to regulate commerce among the several States is not a power to regulate industry within the State." Regulating "activities and relations within the state which affect interstate commerce only indirectly" could lead Congress to "assume control of virtually all the activities of the people to the subversion of the fundamental principle of the Constitution." As had become customary, the liberal majority dissented.[24]

The decisions in *Carter* showed how far the liberal and conservative wings of the Court had separated. In the *Schechter* case of the previous spring, the justices had all agreed that the poultry business of the Schechters had only a slight effect on interstate commerce. But the distinction in *Schechter* was one of degree, not whether the effect was "direct" or "indirect." From the majority opinion on the *Carter* case one could conclude that no matter how enormous the impact, so long as it was indirect, it remained out of reach of federal regulation. The earlier unanimity of the Court had been exploded. The liberal minority could not accept the notion that because the effect was indirect, Congress could not pass legislation regulating the wages and hours in an industry which employed some half-million men digging coal in over thirty states. Without some coal regulation, the liberals anticipated "anarchic riot" in the industry's labor relations. Chief Justice Hughes, who in William Howard Taft's phrase had attempted to "mass the Court" during the early New Deal, now saw his work undone; both left and right had departed from the middle ground he had prepared.[25]

The Carter decision abruptly ended the second attempt to regulate the bituminous industry of America. The commission's work in the fall had been hamstrung by operator resistance, lack of appropriations, and the fears many had that the act might not pass the Court's tests, but now the industry had no stabilization law. William Green, president of the

24. *Carter* v. *Carter Coal Company*, 298 U.S. 317–24ff.
25. Pusey, vol. 2, p. 746; Schlesinger, *Politics of Upheaval*, pp. 477–78. Thomas C. Longin, "Coal, Congress and the Courts: The Bituminous Coal Industry and the New Deal," *West Virginia History* 35 (January, 1974): 101–30, offers a good analysis of the legal questions surrounding the regulation of the coal industry.

American Federation of Labor, suggested that the decision meant that labor would have to rely more on its "economic power." The decision reinforced the miners' loyalties to the president. The *United Mine Workers Journal* noted that the judgment had placed property rights over human rights. "It is a sad commentary on our form of government," said Lewis grimly, "when every decision of the Supreme Court seems designed to fatten capital and starve labor."[26]

The Second Guffey Act: The "Unnecessary Commission"

The Roosevelt administration, the United Mine Workers, and a dwindling number of northern operators quickly prepared a new, stripped-down bill designed to meet the constitutional objections raised by the *Carter* decision. The draftsmen replaced the labor sections with the weak statement that it was the "public policy of the United States that employees shall have the right to organize and bargain collectively through representatives of their own choosing and shall be free from interference, restraint, or coercion of employers." Taking a cue from Chief Justice Hughes, the bill declared that the industry needed price regulation to "remove burdens" from interstate commerce. To avoid delegating power, local coal boards would propose prices that the national coal commission would establish. The tax and fair trade sections from the 1935 act remained.[27]

Despite quick House passage in June, the Senate refused to act on the Guffey bill before adjournment in the summer of 1936. The alliance among Lewis, the northern operators, and the Roosevelt administration solidified. Indeed, the Court's hammer blows against the New Deal and the dramatic legislation of the Second Hundred Days had made the

26. *New York Times*, May 19, 1936; *Literary Digest* 121 (May 30, 1936): 5–6. The *Princeton* (West Virginia) *Observer* celebrated the "overthrow of the Goofy Coal Bill," May 21, 1936; *Coal Age* 41 (July, 1936): 300.

27. John L. Lewis to Frederick A. Delano, May 19, 1936, Delano Papers. Guffey, Solicitor General Stanley Reed, Assistant Attorney General John Dickinson, Charles O'Neill, and John L. Lewis worked on the drafting. *Coal Age* 41 (June, 1936): 260; *New York Times*, May 20, 21, 1936; William Jett Lauck diary, entry for May 22, 1936, Lauck Papers.

president seem to be the only hope for Americans who sought any kind of reform legislation.

Organized labor, in particular, moved into Roosevelt's camp. As a candidate in 1932, Roosevelt had failed to make a single major labor address. Union labor initially viewed his administration skeptically. Section 7a and the union rebirth started some movement to Roosevelt. In 1934 labor lawyer Francis Biddle informed Roosevelt that labor had elected Democrats in Republican Pennsylvania, "particularly in the steel mills and coal mines, where employer domination had been synonymous with Republican control." Social Security, the Wagner Act, the WPA, and the other liberal achievements of the New Deal cemented the alliance. Lewis continued to view the patrician Roosevelt with "the contempt held by a slugger for a boxer," but ewis was booming in 1936 that labor owed Roosevelt a ꝛe debt "that can be liquidated only by casting its solid for him at the coming election."[28]

ꞏis threw the financial resources of the UMW behind ꞏlt's reelection campaign. According to George ꞏ wily Roosevelt perhaps shrewdly extracted more ꞏne workers' chief initially intended to offer, but of the union to the president's campaign— ꞏf million dollars—made the UMW the largest ꞏr. Lewis marshaled the surprisingly effec- Non Partisan League, the political arm of ꞏ, to get out the union vote. Backed by by Democratic conservatives like Al ꞏavis, Roosevelt waged a campaign ꞏen men" against the "malefactors of and exploit the people."[29]

ꞏ, *Politics of Upheaval*, pp. 592–93; Saul ꞏed Biography (New York: G. P. Putnam's ꞏhy Labor Should Support Roosevelt ꞏ 2.

ꞏections of Fifty Crowded Years ꞏ ꞏ; Harold Ickes, *The Secret D*ꞏ ꞏchuster, 1953, vol. 1, p. 2ꞏ William E. Leuchtenbꞏ ꞏ, ed. Arthur M. Schꞏ ꞏo., 1971), vol. 3ꞏ ꞏ94.

ꞏe ꞏty ꞏme and new ꞏt the chief ꞏe any- is this ꞏave this ꞏd by his

"Election of of a Politician ꞏmnly Warn the ꞏ; Carl N. Degler, ꞏation," *Journal of from Washington* ꞏis, p. 177.

The moderate-to-liberal Republican candidate Alfred M. Landon allowed arch conservatives in the Republican Party to portray him as the last defense against the president's march "toward Moscow." In a cruel and unfair jab, Lewis derided Landon as a "bootlicker of plutocracy." Landon could not counter the emotional, class appeal of the president. In the climactic speech of the campaign to a delirious crowd of partisans in Madison Square Garden, Roosevelt declared that the "roll of honor" of his supporters had written on it the "names of millions who never had a chance— men at starvation wages, women in sweatshops, children at looms." In contrast, he stated, the Republican Party had fallen under the hand of plutocrats, the forces of "organized money." These forces, Roosevelt announced, are "unanimous in their hate for me—and I welcome their hatred. I should like to have it said of my first administration that in it the forces of selfishness and of lust for power met their match. I should like to have it said of my second Administration that in it these forces met their master."[30]

The sweeping victory—predicted accurately by Postmaster General James Farley and Joe Guffey—capped the most significant realignment era in twentieth-century American politics. Millions of voters of all classes flocked into the Democratic party to establish its hegemony for the next few years. Disproportionate numbers of new Democrats came from urban areas and from the ranks of the unskilled and semiskilled workers. Labor became a major force in the Democratic coalition. As Lewis's later remarks about the president dining "at labor's table" indicated, the miners expected quid pro quo. He told one newsman: ". . . one fool enough to believe for one instant that we gave money to Roosevelt because we were spellbound by his voice?"[31]

30. Schlesinger, *Politics of Upheaval*, pp. 594, 606; Leuchtenburg, p. 2839.

James A Farley, *Behind the Ballots: The Personal History* (New York: Harcourt, Brace and Co., 1938), p. 323; Lewis, "I Solemnize Industry," *Vital Speeches* 3 (January 15, 1937): 203–20; Political Parties and the Rise of the City: An Interpre*story* 51 (June, 1964): 41–58; Marquis Childs, *I Write* Harper & Bros., 1942), p. 123; Alinsky, *John L. Lewis*

When the new Congress convened, Democratic senators overflowed into the normally Republican right side of the chamber, and southern Democrats in both houses found themselves massively outnumbered by Democrats from other regions.[32] Under such favorable conditions, Lewis quickly cashed in. The Guffey bill shouldered aside an alternative providing for presale price filing and sped through the House without a record vote on March 11. On April 5, 1937, the Senate amended the proposal to place western Kentucky, Illinois, Indiana, and Iowa into a separate price area, changed the tax provisions, and passed the bill, 58 to 15. On April 26 Roosevelt signed the bill into law and appointed the new commissioners.[33]

The new Guffey Act established another National Bituminous Coal Commission (NBCC) of seven members (three "disinterested," two for the miners, and two operator representatives). The twenty-three operator-dominated district boards would propose minimum prices, which the NBCC would establish. In place of the tax-and-rebate system of the first Guffey Act, the new law levied a 1-cent-a-ton tax on all coal sold in the United States and penalized noncode members 19½ cents a ton. A new Consumers' Counsel was created, and a variety of tasks—many the remnants of earlier legislative proposals—were assigned to the NBCC for study.[34]

The act directed the operators in the twenty-three districts to elect district boards. Exempt from the antitrust laws for any actions they took in administering the code promulgated by the commission, board members could only be removed for "inefficiency, willful neglect of duty, or malfeasance in

32. Leuchtenburg, "Election of 1936," p. 2842; Leuchtenburg, *Franklin D. Roosevelt and the New Deal,* p. 196.
33. Southern coal operators tried to supplant the second Guffey bill with a price-filing plan but failed. Untitled Memorandum, January 27, 1937; Charles E. Smith to Van Bittner, November 10, 1936, Box 22, Smith Papers; *Coal Age* 42 (March, 1936): 138; (April, 1936): 171–73; *New York Times,* March 12, April 2, 6, 1937; *Pittsburgh Press,* April 9, 1937.
34. *Newsweek,* May 15, 1937, p. 14. Undated, untitled newspaper clipping; Jouett Shouse to Charles E. Smith, March 16, 1932, Box 8, Smith Papers; *Coal Age* 42 (June, 1937): 293; Baker, p. 111, note 67; Bituminous Coal Act of 1937, 50 Stat. 72 (1937).

office." The boards were to propose minimum prices based on the "average cost of production" in their district.[35]

Statistical bureaus in each district were to aid the boards in arriving at this "average cost of production." Code members had to submit copies to the statistical bureaus of "all contracts, invoices, credit memoranda, and any other information concerning the cost, sale, and distribution of bituminous coal as the commission might require." With this cost information, correlated by the statistical bureaus, the district boards would propose the minimum prices for the commission to review and establish as legal minimums in ten price areas.

When different railroad rates or other factors caused prices submitted by one district on a particular type of coal to be either higher or lower than those from another district which traditionally shipped into the same minimum-price area, the commission could order the districts to coordinate their prices so that the prices returned to the producers an average realization per ton equal to the "weighted average cost of production" in the price area. This system protected high-cost mines with poor locations and made the law difficult to administer. To avoid the interdistrict competition which had weakened the NRA attempt to fix minimums, the 1937 Coal Act created a byzantine system designed eventually to allow the commission to establish some four hundred and fifty thousand prices. Since the commission would regulate every interstate coal sale in the nation, this minimum-pricing system of the Guffey-Vinson Act required colossal effort by the commission and thorough cooperation from the operators.[36]

This last attempt at business-government cooperation confronted the same regional splits and economic difficulties which had plagued the industry since the beginning of the century.[37] The Supreme Court reversed itself dramatically on its interpretation of the commerce power and specifically up-

35. Bituminous Coal Act, Sections 1–14; Fisher and James, pp. 307–16.
36. Bituminous Coal Act of 1937, Sections 2, 4–6, 10–19; Baker, pp. 67–80; Fisher and James, chapter 3; Interuniversity Case Program, *The Consumers' Counsel and the National Bituminous Coal Commission, 1937–1938* (Washington, D.C.: Committee on Public Administration Cases, 1950), pp. 40, 66–67.
37. Baker, chapter 6; Edward W. Carter, "Price Fixing in the Bituminous Coal Industry," *Annals* 193 (September, 1937): 123–29.

held the law in the *Sunshine Anthracite* case. But competitive and political pressures doomed the new Guffey Act.[38] For the first months the commissioners embroiled themselves in patronage squabbles as they tried to fill the three hundred positions created by the law. They accused each other of using the appointments to further their individual political ambitions. Lewis wrote Roosevelt in November, 1937, that the NBCC was in a "hell of a state." In an attempt to promote harmony, Chairman Charles F. Hosford submitted his resignation. The president instructed him to stay on and told the commission to "quit wrangling and fix prices."[39]

To fix minimums, operators in district board meetings had first to agree on how to classify coal of various types, sizes, and quality. They could not. At a commission hearing in July, operators played Uriah Heep, explaining how poor their region's coal was to get a lower classification and price. Some operators wanted to continue the industry's traditional policy of classifying coal on the basis of the use to which the consumer put it. In the past, operators had granted the railroads and the utilities special prices on steam coals. This policy required the industry to charge higher prices to other consumers. Henry Warrum, general counsel for the UMW, asked the operators at the hearing how long they were going to allow the railroads to drive prices down below costs. Chairman Hosford listened to arguments back and forth. On August 16, 1937, the commission refused to allow the district boards to classify coals according to use. The NBCC then issued a ruling establishing classification standards.[40]

38. *Sunshine Anthracite Coal Co.* v. *Adkins*, 310 U.S. 381 (1940); *National Labor Relations Board* v. *Jones & Laughlin Steel Corporation*, 301 U.S. 1 (1937); *Time*, May 13, 1940, p. 83; *New York Times*, February 25, 1937; *Coal Age* 42 (May, 1937): 181–82.

39. Charles E. Smith to Charles Hosford, November 24, 1937; Charles E. Smith Memorandum, May 29, 1937; R. H. Bailey, Jr., to Smith, May 24, 1937; Harry Brown to Smith, August 18, 1937, Smith to R. H. Bailey, Jr., August 24, 1937, all in Boxes 23–24, Smith Papers; James Rowe, Jr., to Marvin McIntyre, November 21, 1938, OF 1732, Box 1, Roosevelt Papers. Lewis quoted by Drew Pearson, "Washington Daily Merry-Go-Round," *Washington Herald*, November 14, 1937; Roosevelt quoted by Baker, p. 109.

40. Charles F. Hosford to Marvin McIntyre, August 17, 1937, OF 1732, Roosevelt Papers; Hosford to Charles E. Smith, October 30, 1937, Box 24, Smith Papers; *Coal Age* 42 (August, 1937): 373; Baker, pp. 143–48; Fisher and James, chapters 5–8.

But the districts began to fight over coordinating prices in market areas. The three district boards in Pennsylvania could not coordinate prices. The commission summoned the members and arbitrated the dispute so that the Pennsylvania boards could begin to coordinate their prices with the other boards which sold in the same minimum-price areas. In mid-October the districts found they could not coordinate. "I want to serve notice upon this industry right now," Hosford informed the industry, "and upon my fellow Commissioners, that I am perfectly willing to have the Commission undertake the job of proposing and establishing these prices without any further help from the industry." The following day, neglecting the required procedural steps and hearings, the commission began to coordinate prices without the advice of the industrialists. On November 30, 1937, the commission itself promulgated minimums to be effective two weeks later.[41]

Large consumers, including the City of Cleveland and two hundred railroads, immediately sought injunctions against the new prices on due-process grounds. The commission retorted that public hearings at which "every one of 500,000 prices could be assailed, would require months." The consumers got their injunctions at various circuit courts of appeals. John L. Lewis and fifteen major operators agreed that the injunctions nullified the established prices. Rather than be left with an "empty shell," the commission revoked the minimum prices as of February 25, 1938.[42] Following a round-table discussion of representatives of the district boards in March, the commission pledged it would hold public hearings at each stage, including a final hearing on the coordinated minimum prices.

In April, 1938, a full year after the act's passage, the commission ordered the district boards once again to determine

41. Baker, pp. 151–54; Fisher and James, pp. 50–52. Although the NBCC lacked the legal power, it issued a subpoena to District 15 to compel them to file prices; the threat worked.

42. Interuniversity Case Program, pp. 90–95; *Business Week*, February 12, 1938, p. 14; February 19, 1938, p. 29; Roosevelt to John E. Miller, February 17, 1938, OF 1732, Box 1, Roosevelt Papers; *New York Times*, February 24, 1938; Baker, pp. 159–64; Fisher and James, pp. 52–54.

weighted average costs for their districts. The commission then held public hearings on the cost determinations in Denver, Washington, D.C., and the Southwest. On the basis of the cost figures, the commission ordered the districts to submit minimum prices by August, 1938. Hearings on the proposed prices took until December, 1938, and ran to nearly eight thousand pages. The commission next required the district boards to coordinate prices by February 15, 1939, then changed the deadline to March 15. The job proved impossible. The commission assumed the task a second time, but with public hearings.[43]

As it struggled to complete its price list during the summer of 1939—two years after its creation—Roosevelt sacked the commission. Weary of the delay and the reports of dissension within the commission, Roosevelt transferred the functions of the commission into Harold Ickes's Department of the Interior as part of the Reorganization Act of 1939. "Experience," said the president, "has shown that direct administration will be cheaper, better, and more effective than through the cumbersome medium of an unnecessary commission." Indeed, the NBCC came to symbolize the ineptitude and waste of New Deal bureaucracy. The NBCC, Ickes mused privately, "has been one of the worst-managed and most malodorous organizations in the entire national administration."[44]

The Department of the Interior's Bituminous Coal Division carried on the price-fixing program thereafter. Harold Ickes dismantled half of the old commission's statistical offices and fired one-third of the field personnel and fifty-four persons from the Washington offices. The division's final hearings took over one hundred days and ran into early 1940. On the basis of the hearings, which were "one of the most exhaustive inquiries ever conducted" on one industry's affairs, the division issued proposed findings and, following further hearings and much delay, promulgated mini-

43. Baker, chapter 6; Fisher and James, chapter 8.
44. Parker, pp. 146–47; Interuniversity Case Program, pp. 115–18; *Business Week*, April 6, 1940, p. 17; Baker, p. 121. Ickes to Roosevelt, July 17, October 2, 1939; Roosevelt to Ickes, October 13, 1939, all in OF 6AA, Roosevelt Papers; Ickes, *Secret Diary*, vol. 2, pp. 630–31.

mum prices effective October, 1940—three and one-half years after the passage of the Guffey-Vinson Act.[45]

The coal division's achievement never received a fair trial. As the nation began to revive economically with the beginning of the war economy in 1940, coal production jumped by 15 percent over 1939 and by another 11 percent between 1940 and 1941. Average per-ton prices went from $1.84 in 1939 to $2.36 in 1942. The division revoked 139 memberships where producers violated prices, it enforced its minimums, and it altered some prices on petition. But growing demand removed the pressure on prices. Consumers charged, in fact, that coal prices were becoming "excessive and oppressive." Consumer pressure for maximum prices finally led Roosevelt to have the Office of Price Administration, with statistical assistance from the coal division, establish maximum prices in May, 1942. Congress refused to reenact the coal act when it reached the end of its legal life during the war.[46]

Conclusions

The New Deal's attempts at industrial stabilization of the coal industry through business-government cooperation had ended on a sour note. The Guffey-Vinson Act failed to set forth adequate criteria for determining costs, classifying coals, or establishing minimum price schedules in the bituminous industry. Aimed at building a floor under prices so that the industry could pay adequate wages, an imprecise law gave the district boards too much latitude, which inevitably led to extensive bargaining and tended to freeze the economic condition of the industry.[47]

The politically conscious commissioners compounded the difficulties through maladroit administration. The Interior Department's coal division did somewhat better, but like

45. Parker, p. 147; Baker, pp. 133–34, 175–78; *Newsweek*, August 12, 1940, p. 32.

46. See Table 4. Edwin C. Witte, "What the War is Doing to Us," *Review of Politics* 5 (January, 1943): 3–25; Fisher and James, pp. 64, 451–78; Longin, "Coal, Congress and the Courts," p. 126.

47. Fisher and James, pp. 307–49.

their NRA predecessors, the bureaucrats who tried to stabilize the bituminous industry under the Guffey-Vinson Act were fighting enormous obstacles. Establishing "fair" prices required taking into account the relative market values of differing kinds, qualities, and sizes of coal and different producers from diverse regions, each with its own railroad and wage rates, all competing in various markets and selling to consumers who sought to use the coals for distinct purposes. The reality of bituminous coal pricing, Frederick C. Mills has written, is like an iceberg. An economist sees only the tip in the spot quotations for various markets. Nine-tenths lies invisible beneath the sea.[48]

If the war had not intervened, would the Guffey-Vinson Act have stabilized the bituminous industry? It probably would have reduced price cutting, as NRA had done briefly. It might have helped the union maintain wages and working conditions and could have reduced strikes. Given better prices, the industry might have earned more. But the Bituminous Coal Commission was a bureaucratic behemoth, unable to adapt to economic changes. Basing prices on costs would have driven prices up in recession and depression as production fell and fixed costs became a higher percentage of total costs, thus maintaining overdevelopment in an industry plagued by its excess capacity. By curing the immediate symptoms, the law might have weakened the "patient's ability to rally from his fundamental disease."[49]

Regardless of the briefly stable prices achieved under NRA and the incredible efforts of the coal division personnel under the Guffey-Vinson Act to devise a system for establishing minimum prices, neither probably would have succeeded in the long run. Competitive operators could always find loopholes in any code of fair competition. Regions would always defend themselves, and by altering their markets with cost data, minor differences in contract terms could undermine any "price." Indeed, a price was not a number, but a value as perceived by a buyer, and in the coal

48. Quoted in ibid., p. vii, see also pp. 114–281, 402–40.
49. Ibid., pp. 439–44; Longin, "Coal, Congress and the Courts," p. 130.

industry it encompassed more variables than any law or agency could ever regulate.

Regulation based on past industry performance was an inadequate program for the depressed coal industry. The supporters of the NRA and the Guffey acts mistakenly assumed that raising prices and wages in the industry would somehow revive it. Raising prices in the coal industry reduced the industry's aggregate operating deficit, but it encouraged retention of high-cost mines and the opening of marginal mines. The New Dealers thus perpetuated the industry's basic problem of excess capacity. As the complex history of the Guffey acts makes clear, fixing minimum prices for bituminous coal became an enterprise of monumental bureaucratic waste.

Raising wages improved the lot of the working miner in the short run, and—given the poverty and immobility of mine labor during the 1930's—this improvement was a valuable, short-run social and economic gain. Increased wages encouraged mechanization, particularly in the Midwest, and made the industry more efficient, but they led in the long run to a reduction in the labor force. Neither the union, the operators, nor the New Dealers had a solution for the chronic poverty in the Appalachian region. Even special-interest legislation failed the fragmented industry.

9

The Politics of Soft Coal

THE PRECEDING HISTORY of the politics of soft coal from the First World War through the New Deal gives clear answers to the three clusters of questions discussed in the opening chapter. The war experience did not provide the New Dealers regulating coal with a usable past. Industrial self-government also failed in both war and depression. Most important, the politics of soft coal is best understood through a structural approach, rather than in terms of government regulating business in the "public interest," or industry controlling or co-opting governmental agencies.

The Analogue of War

The New Dealers found that without the First World War's expanding economy and centralized controls, the application of war-inspired techniques did not work without the war prosperity. The First World War profits greased the wheels of Fuel Administration regulation. Harry Garfield's Fuel Administration directed the booming production, set effective maximum prices, moderated labor disputes, and brought the industry the stabilization and order that it could not achieve on its own. Hugh Johnson's NRA helped the operators to establish temporarily effective prices and the United Mine Workers to raise wages, improve working conditions, and dramatically reduce the North-South wage differential. But because operator cooperation during the Great Depression came not out of profits but out of desperation, it could not last. The southern wing of the industry, allied with a variety of other operators, fought the NRA throughout its history and eventually had it declared unconstitutional. Regional competition undermined the stabilization programs of the two Guffey acts.

Prosperity and depression produced contrasting labor-management relations. During the war the UMW extracted what it could from the prosperous operators. Patriotism and increased wages worked against strikes and militancy during the war. Indeed, Garfield carefully supported moderate leaders like John P. White to preempt what he called the "radical elements" among the workers. The depression years changed union attitudes. Aggressive leadership by John L. Lewis, abetted by Section 7a of the NIRA, launched the resurgence of a militant union. Local strikes and the threat of a nationwide shutdown helped force the industry under a code and created the nationwide wage scale which cut the North-South wage differentials. The UMW grew during the war because the government sought to prevent strikes and encouraged management to bargain. Moreover, during war, profiting firms could afford union wages. During the depression the UMW quadrupled because it launched a powerful organization campaign. Something of a decorative appendage to the war mobilization program, the UMW during the New Deal became the crucial agent in improving wages, hours, and working conditions. It literally forced the enactment of the two Guffey acts.

The war and depression eras contrasted sharply. The war mobilization came on the heels of the Progressive Era and had a strong antitrust, antibusiness tone. During the war, antitrust progressives like William Colver and Atlee Pomerene sought to bring the "coal barons" to heel. Only after the winter crisis of 1917–18 did the coal industrialists influence the Fuel Administration. Newton Baker and President Wilson condemned the industry-dominated Committee on Coal Production for its three-dollar maximum price. Coal regulation during the First World War thus represented in a real sense a "flowering of progressivism."[1]

The ethos, spirit, and direction of the New Deal coal regulation differed fundamentally from those of the Fuel Administration. The ebullient and unflagging Hugh S. Johnson replaced the reserved, somewhat austere but forceful Harry Garfield. Interregional conferences of operators trying to protect their market areas' prices superceded the centralized price fixing done by the Fuel Administration's engineers' committee. The Fuel Administration, a highly structured administrative agency, gave way to the NRA, a central bureaucracy which tried to cajole and coerce disputatious coal operators into effective regional self-government. In part because depressed conditions required it, but also because of the attitude of President Roosevelt, the NRA set out— quite unlike the war agencies—to improve the conditions of the underprivileged and unemployed. For this reason, among others, the NRA involved the UMW in a much larger role than did the Fuel Administration.

As was true for the Progressive Era generally, the war mobilization programs had a much stronger moral tone than did the New Deal.[2] Wilsonian rhetoric, reinforced by the Committee on Public Information, made saving coal seem to be a matter of protecting the lives of American soldiers in the trenches. Although Hugh Johnson tried to invoke the same kind of spirit, patriotism came easier in prosperity than in depression; rather shortly after the ballyhoo of the NRA parades, coal operators settled back into their struggle for corporate and regional advantages. The more flexible, decentralized approach of the NRA meant that the outright

1. Allen F. Davis, "Welfare, Reform and World War I," *American Quarterly* 19 (Fall, 1967), reprinted as "The Flowering of Progressivism" in *The Impact of World War I*, ed. Arthur S. Link (New York: Harper and Row, Publishers, 1969), pp. 44–56. See also Robert D. Cuff, "Herbert Hoover, the Ideology of Voluntarism and War Organization during the Great War," *Journal of American History* 54 (September, 1977): 358–72.

2. Richard Hofstadter, *The Age of Reform: From Bryan to F.D.R.* (New York: Vintage Books, 1955; originally published by Alfred A. Knopf, 1954), chapter 7; Otis L. Graham, Jr., *An Encore for Reform: The Old Progressives and the New Deal* (New York: Oxford University Press, 1967), chapter 2.

withdrawals of western Pennsylvania and western Kentucky from the code system were possible and tolerated without great moral outrage. No such actions would have been suffered by war bureaucrats.

Thus, although the war years provided the New Dealers with beneficial experience, the different economic conditions dictated a reversal of objectives, techniques, and styles. Each experiment in business-government cooperation succeeded in some ways, but war and depression were economic opposites. Mobilization could not provide a truly usable past for the coal programs of the New Deal. The more moralistic, structured Fuel Administration told coal men in a boom economy what they must do to win the war. The more flexible, decentralized NRA, supported by the powerful UMW it spawned, struggled to convince coal men in a depressed economy that they should cooperate for their survival.

Industrial Self-Government

Tried under the Committee on Coal Production, industrial self-government never had a chance to work in 1917. The operators' three-dollar maximum price angered Woodrow Wilson and ended a voluntary program. When reapplied to coal in the 1930's, industrial self-government succeeded only briefly. Influenced by early and massive propaganda and pressed by the UMW into a national contract, the fragmented coal industry agreed to a code of fair competition. Hammered out in the hearings and meetings of late summer, 1933, the bituminous coal code tried to balance the interests of the many producing areas, each with its different coal deposits, costs, mining conditions, freight rates, and markets. It declared illegal many vicious and common competitive trade practices. By incorporating the Appalachian agreements, it reversed the downward wage spiral and improved hours and working conditions. And despite the code's purposely vague price-fixing sections and the gyrating NRA price policy, operators established enough monopoly control to halt price cutting briefly.

Regional and private competitiveness, however, soon destroyed industrial self-government. The NRA code gradually collapsed in midwinter 1934–35, never to be revived because of the age-old hostility among different mine fields and particularly between the union operators of the old Central Competitive Field and the non-unionists south of the Ohio River. Regional fights doomed the Guffey-Snyder Act even before the Supreme Court declared it invalid. The same forces, plus revived consumer lobbying, delayed the operation of the Guffey-Vinson Act until World War II, when it was no longer necessary.

Ironically, the United Mine Workers, not the industrialists, provided the great impetus behind industrial self-government in coal. Following his abandonment of the traditional "business unionism" policies of the Gompers era, John L. Lewis marshaled his legal staff to bring about legislative relief for his union, the miners, and the industry. His legislative proposals prepared the way for NRA, for Section 7a, and for the Guffey acts. Indeed, he literally blackmailed the Roosevelt administration into supporting the Guffey-Snyder Act. The resurgence of the UMW and the Appalachian Agreements of 1933 and 1934 created the cost base which compelled operators to cooperate on minimum prices. But although the revived UMW could exercise its muscle and win passage of the Guffey acts, it could not increase demand for coal. John L. Lewis and his union became a real force in the 1930's, but his union proposals for stabilization of the coal industry were no long-term solution.

The Question of Control

During both the First World War and the New Deal, the coal industry proved too disunited to manipulate government for its own economic benefit. Sometimes thought to be a cohesive whole, even the "coal trust," as during the dramatic inflation of 1917, when senators denounced the "coal barons" for having caused high prices through some kind of monopoly control or conspiracy, the coal industry proved too fragmented to exploit government regulation, much less co-opt any regulatory agency. Except for 1917–18, the industry suffered from fierce competition and in most markets from excess capacity; large, powerful industrial purchasers operating in buyers' markets brought low prices. Attempts before World War I to stabilize the industry by expanding foreign markets, by pressing the Interstate Commerce Commission to restructure freight rates, by creating a national trade association, by regional sales agencies or open price associations, by direct price agreements, and by mergers all failed.

Even the cohesion the industry achieved under the Fuel Administration failed to match the unity and political effectiveness of other industries. The Fuel Administration coordinated the coal industry and with operator assistance brought stabilization, but the industry never exploited or controlled the Fuel Administration. On the contrary, the Fuel Administration held down the industry's prices and profits.

Despite the industry's good beginning toward rationalization under the Fuel Administration, from 1926 onward, weakening demand eroded prices, and the industry splintered north and south, union and non-union. The divided operators watched profits evaporate in internecine competition. The National Coal Association—a product of the war—effectively prevented regulatory legislation early in the 1920's but it did not unite the industry behind a plan to stabilize prices during the Great Depression. Commerce Secretary Hoover's plans fell victim to the 1922 strike; other proposals, including a revival of joint sales agencies, met operator disinterest.

Required to draft an NRA code of fair competition, the coal operators balked, submitted more than two dozen regional codes, and dragged out the code-drafting process through the summer of 1933. Once forced under a single code, the regions briefly evolved a fairly satisfactory price-stabilization system based on market areas, but regional and corporate rivalries began to break down these arrangements in 1934. The union and the northern operators fought to enact a stabilization bill for soft coal, but southern coal men resisted and then convinced the Supreme Court to declare the Guffey-Snyder Act unconstitutional. Competitive rivalries and poor administration delayed operation of the Guffey-Vinson Act until 1940, when the war obviated the need for minimum prices.

During both the New Deal and the First World War, politicians attacked the industry, particularly its more reactionary members, in the name of the "public interest." John F. Kennedy's allusions to the parentage of businessmen would probably accurately characterize some of the worst antisocial operators. But unlike the monolithic oil or steel industry, the soft coal industry lacked the cohesion to control government or its regulatory agencies enough to exploit the "public." Coal men directly influenced the Fuel Administration and ran their NRA code authorities. The Central Field operators allied with the UMW to get two special pieces of legislation to protect their particular interests, but they did not control the New Deal commissions. The one force in the politics of soft coal that could powerfully affect government during the New Deal was, of course, the UMW. But the measures it backed were themselves short-sighted and eventually contributed to reducing the size of the union's membership over time.

The politics of soft coal in war and depression, then, was not a story of regulation in the "public interest," although some saw it as such. Nor was it a tale of businessmen controlling or co-opting government agencies. The politics of soft coal was rather a story of a splintered industry unable to find a means to save itself through "industrial self-government." In applying the analogy of war, the New Dealers opened the way for a midwestern operator-union alliance that produced the Guffey acts. But the failure of this special-interest legislation showed how significantly structural and regional divisions within the industry affected the politics of soft coal.

Bibliographical Essay

TO LIST THE numerous sources already cited in the text would be a duplication. Moreover, since many of the works cited are old friends to any historian working in twentieth-century American history, they need not be singled out for special mention. Of particular help in general were two journals, *Coal Age* and the *United Mine Workers Journal*. Each has its bias, but each maintained writers who covered the day-to-day events of the industry and had editors who analyzed them. Two general economic studies of the coal industry, Carroll L. Christenson, *Economic Redevelopment in Bituminous Coal* (Cambridge, Mass.: Harvard University Press, 1962) and Reed Moyer, *Competition in the Midwestern Coal Industry*, Harvard Economic Studies, vol. 122 (Cambridge, Mass.: Harvard University Press, 1964), while written about a later time period, are indispensable to a grasp of the economics of bituminous coal. See also Harold Barger and S. H. Schurr, *The Mining Industries, 1889–1939* (New York: National Bureau of Economic Research, 1944). A good history of the industry's relationship with government based on secondary sources is Glen Lawhon Parker, *The Coal Industry: A Study in Social Control* (Washington, D.C.: American Council on Public Affairs, 1940).

The War Years

The standard works by Melvin I. Urofsky, Robert Wiebe, Gabriel Kolko, James Weinstein, and Robert D. Cuff treat business-government relations during the Progressive Era and World War. See also Paul A. C. Koistinen, "The 'Industrial-Military Complex' in Historical Perspective," *Business History Review* 41 (Winter,

1967): 379–403; K. Austin Kerr, *American Railroad Politics 1914–1920: Rates, Wages, and Efficiency* (Pittsburgh: University of Pittsburgh Press, 1968); Albro Martin, *Enterprise Denied: Origins of the Decline of American Railroads, 1897–1917* (New York: Columbia University Press, 1971); William F. Willoughby, *Government Organization in War-Time and After* (New York: D. Appleton-Century Co., 1919); and Charles R. Van Hise, *Conservation and Regulation in the United States During the World War* (Washington, D.C.: Government Printing Office, 1917).

On the coal industry during the Progressive Era and the war, good starting points are William Graebner, "Great Expectations: The Search for Order in Bituminous Coal, 1890–1917," *Business History Review* 48 (Spring, 1974): 49–73, Graebner, *Coal Mining Safety in the Progressive Period: The Political Economy of Reform* (Lexington: The University Press of Kentucky, 1975), and Robert Cuff, "Harry Garfield, the Fuel Administration, and the Search for a Cooperative Order during World War I," *American Quarterly* 30 (Spring, 1978): 39–53. From there one must turn to the extensive collection of documents in the Records of the United States Fuel Administration (Record Group 67), National Archives. The careful, if adulatory, reports of the United States Fuel Administration bureaucrats, particularly *Final Report of the United States Fuel Administrator, 1917–1919*, the *Report of the Administrative Division, 1917–1919*, the *Report of the Distributive Division, 1918–1919*, and the *Report of the Engineers Committee, 1918–1919*, all published by the government in 1919 and 1920, help establish the chronology and give context.

Three other government publications, U.S. Federal Trade Commission, *Cost Reports, Coal, Nos. 1–9* (1919), the FTC's *Preliminary Report on Investment in Soft-Coal Mining* (1922), and Paul W. Garrett, *History of Government Control of Prices* (Washington, D.C.: Government Printing Office, 1920), are invaluable in determining effectiveness of price regulation. A fourth government publication, U.S., Congress, Senate, Committee on Interstate Commerce, *Price Regulation of Coal and Other Commodities, Hearing on S. 2354 and S.J. Res. 77*, 65th Cong., 1st sess., 1917, is extremely helpful in untangling the confusions surrounding the failure of the Council of National Defense's coal production committee to work harmoniously with the Wilson administration.

All of these government sources must be read in connection with the papers of Woodrow Wilson, Harry A. Garfield, William Gibbs McAdoo, Josephus Daniels, and Newton Baker in the

Manuscript Division, Library of Congress; the investigations cited in the Department of Justice Central Files; and the Federal Trade Commission Records. The microfilming of the Wilson Papers and the publication of an index to these voluminous records are of immense importance to historical scholarship.

Between the War and the New Deal

The standard works on these years by William E. Leuchtenburg, Robert Murray, George Soule, Paul Carter, and Irving Bernstein are indispensable. But except for Bernstein, they contain little on the coal industry. Ellis W. Hawley's "Secretary Hoover and the Bituminous Coal Problem, 1921–1928," *Business History Review* 42 (Autumn, 1968): 247–70, and his "Herbert Hoover, the Commerce Secretariat, and the Vision of an 'Associative State,' 1921–1928," *Journal of American History* 61 (June, 1974): 116–40, are most helpful on Hoover's efforts to stabilize the industry. Louis D. Galambos, *Competition & Cooperation: The Emergence of a National Trade Association* (Baltimore: Johns Hopkins University Press, 1966); Robert F. Himmelberg, *The Origins of the National Recovery Administration* (New York: Fordham University Press, 1976); and Gerald D. Nash, *United States Oil Policy, 1890–1964* (Pittsburgh: University of Pittsburgh Press, 1968), are very helpful on trade associations and government policies during the 1920's.

On the relationships between the New Deal and the war mobilization, see Leuchtenburg, "The New Deal and the Analogue of War," in *Change and Continuity in Twentieth-Century America*, ed. John Braeman et al. (New York: Harper and Row, 1964), pp. 81–144; Nash, "Experiments in Industrial Mobilization: W.I.B. and N.R.A.," *Mid-America* 45 (January, 1963): 157–74; and Nash, "Franklin D. Roosevelt and Labor: The World War I Origins of Early New Deal Policy," *Labor History* 1 (Winter, 1960): 39–52.

Many of the hearings held on special legislation for the industry are helpful, but they are too numerous to be cited here. The U.S., Coal Commission, *Report*, 5 Parts, 68th Cong., 2d sess., 1925, while ineffectual at the time, contains extremely helpful data. On the labor situation, see U.S., Congress, Senate, *Report No. 457, West Virginia Coal Fields*, 67th Cong., 2d sess., 1922, and U.S. Congress, Senate, Committee on Interstate Commerce, *Conditions in the Coal Fields of Pennsylvania, West Virginia, and Ohio, pursuant to S. Res. 105*, 70th Cong., 1st sess., 1928.

Three articles deserve special mention as providing perceptive analysis: F. G. Tyron, "Effect of Competitive Conditions on Labor Relations," *Annals of the American Academy* 111 (June, 1924): 83–95; Tyron, "The Irregular Operation of the Bituminous Coal Industry," *American Economic Review* 11 (March, 1921): 57–73; and M. B. Hammond, "The Coal Commission Reports and the Coal Situation," *Quarterly Journal of Economics* 38 (1924): 541–81.

The Coal Industry during the New Deal

In addition to the standard works by Hugh S. Johnson, William E. Leuchtenburg, Arthur M. Schlesinger, Jr., Frank Freidel, James MacGregor Burns, Charles F. Roos, George B. Galloway, Leverett S. Lyon et al., Sidney Fine, and Ellis W. Hawley, a new biography of John L. Lewis by Melvyn Dubofsky and Warren Van Tine (New York: Quadrangle/New York Times Book Co., 1977), proved very useful. See also Thomas C. Longin, "Coal, Congress and the Courts: The Bituminous Coal Industry and the New Deal," *West Virginia History* 35 (January, 1974): 101–30.

Unpublished government sources helpful in understanding the bituminous coal code's operation are U.S. Special Industrial Recovery Board, "Proceedings," Washington, D.C. (mimeographed copy in the Columbia University School of Business Library), 1933–34; and three studies in Record Group 9, National Archives: Newell W. Roberts, et al., "History of the Code of Fair Competition for the Bituminous Coal Industry" (typescript), Records of the Division of Review, Code Histories for Industries under Approved Codes; Thomas S. Hogan, "A Short History of the Activities of the Coal Labor Boards" (typescript), Consolidated Files of Typescript Studies Prepared by the Division of Review; and Andrew Pangrace, "Preliminary Abstract of Work on the Labor Compliance Activities of the Bituminous Coal Labor Boards" (typescript), Records of the NRA Organization Studies Section, Preliminary Drafts of Reports.

The voluminous Records of the National Recovery Administration in the National Archives contain the raw material necessary for any study of a code of fair competition. Less organized are the Records of the National Bituminous Coal Commission. Other manuscript collections helpful in this connection were the collections in the Franklin D. Roosevelt Library, Hyde Park, New York; the United Mine Workers collection in the AFL-CIO Library, Washington, D.C.; the John Brophy Papers, Catholic University of America, Washington, D.C., which are also useful on the 1920's; the William Jett Lauck Papers, University of Virginia Library, Charlottesville, Virginia; the Gifford Pinchot and Donald Richberg Papers, Manuscript Division, Library of Congress; the Charles E. Smith and Smokeless Operators Association Papers in the West Virginia University Library, Morgantown, West Virginia; and the Edward A. Wieck Papers, Labor History Archives, Wayne State University, Detroit, Michigan.

In particular, William Jett Lauck's diary (typescript) and his "Docket—United Mine Workers" (typescript) in his papers were of great assistance in the legislative history of the Guffey acts. On the rise of the UMW, Wieck's "The Miners' Union in the Steel Industry" (typescript) and his binders of interviews with UMW organizers and coal operators are excellent, if partisan.

Four mimeographed studies in the U.S. National Recovery Administration Work Materials series were outstanding. Primary, of course, is "Economic Study of the Bituminous Coal Industry under Free Competition and Code Regulation," by F. E. Berquist et al., Work Materials No. 69 (1936). Helpful were "Code Authorities and Their Part in the Administration of NIRA," Work Materials No. 46 (1935); "Production, Prices, Employment and Payrolls in Industry, Agriculture, and Railway Transportation," Work Materials No. 15 (1935); and "Minimum Price Regulation under Codes of Fair Competition," by Saul Nelson, Work Materials No. 56 (1936). One published hearing deserves note: U.S., Congress, Senate, Committee on Finance, *Investigation of the National Recovery Administration, Hearings pursuant to S. Res. 79*, Six Parts, 1935.

In addition, several economic studies are good background to the study of the coal industry during the New Deal. Waldo E. Fisher, *Economic Consequences of the Seven-Hour Day and Wage Changes in the Bituminous Coal Industry* (Philadelphia: University of Pennsylvania Press, 1939), analyzes the impact of changes wrought by the New Deal and the labor contracts of the 1930's. With Charles E. James, Waldo Fisher wrote the definitive work on the price program of the Guffey acts, *Minimum Price Fixing in the Bituminous Coal Industry* (Princeton: Princeton University Press, 1955). This should be read with Ralph Hillis Baker, *The National Bituminous Coal Commission* (Baltimore: Johns Hopkins University Press, 1941). W. G. Fritz and T. A. Veenstra, *Regional Shifts in the Bituminous Coal Industry* (Pittsburgh: University of Pittsburgh Bureau of Business Research, 1935), explains much about the economic conditions underlying the problems the industry faced in the 1930's. Joseph T. Lambie, *From Mine to Market: The History of Coal Transportation on the Norfolk and Western Railway* (New York: New York University Press, 1954), explains the interrelationship between railroad transportation and coal mining.

Index

Advisory Commission of the Council of National Defense, 35, 37
Agricultural Adjustment Act, 142
Allocation plan proposed by operators, 207
Analogue of war: employed by New Deal, 138, 139–40, 195, 201–2
Anthracite strike of 1902, 23
Appalachian Agreements: 1933, 183; 1934, 163–64, 185–86; 1935, 221
Appalachian Coals, Inc., 120
Appalachian producing area, 13

Baer, George, 23
Baker, Newton D., 18, 35, 36, 44–45, 65–66, 97
Barnes, Charles, 179–81
Baruch, Bernard M., 10, 36–37, 61
Berle, Adolph A., 142–45
Bittner, Van, 169
Bituminous Coal Commission. See United States Coal Commission
Bituminous Coal Labor Boards of NRA, 177, 178–82
Black, Hugo S., 142
Borah, William, 168, 205
Brophy, John, 171
Business-government cooperation during World War I: planned, 35; failed, 45, 50; reactivated, 71, 80–81; collapsed again, 108
Business-government relations: public-interest approach, 2, 12, 51, 94, 134, 146, 163–64, 243–45; capture approach, 3, 12, 51, 58, 92–93, 134, 163–64, 200–201, 243–45; structural approach, 4–5, 12, 92–94, 163–64, 219, 239–45; cooperative goal, 5; after 1919 panic, 27; wartime, 93; Garfield on, 94, 98–101; self-regulation approach, 145; Franklin Roosevelt on, 145–46

Calder, William M., 110
Callbreath, James, 27
Captive mines, 22, 154–56, 173–75
Central (Competitive) Field, 13; 1898 wage agreement, 2, 14, 26; organizing difficulties, 1880–1917, 19, 27, 28; Cleveland Agreement, 1922, 116; Jacksonville Agreement, 1924, 118–19; propose coal bill, 1933, 140; and freight rates, 193; alliance with UMW, see United Mine Worker–northern operator alliance
Coal Distribution Committee, 1922, 114
Coal Division, Department of the Interior, 225–26
Coal panic of 1920, 107
"Coal Trust": Wilsonians and, 18–19; and anthracite strike of 1902, 23; assailed or alleged, 32, 43, 46, 96; and reality, 245

Code of Fair Competition of NRA: structure, 158, 160; and collective bargaining, 171–82; wages, 182–89; compromises in, 197–98; price provisions, 198–99; code authorities and, 198–99; administrative weaknesses, 199

Coiner, Charles T., 148

Collective bargaining: and 1920's, 14, 114, 116, 119–20, 121; in 1917, 81, 82; and Franklin Roosevelt in 1932, 137; and Section 7a of NIRA, 144. *See also* United Mine Workers; Wages

Colver, William B., 32, 34, 54

Committee on Coal Production, 18, 37, 38, 39, 41, 42, 44

Competitive structure of coal industry: Temporary National Economic Committee on, 23; overcome under NRA, 202–4. *See also* Structure of coal industry

Conflict of interest, 48, 72

Consumers of coal, 22, 87, 89, 204–7, 244

Coolidge, Calvin, 117

Cooperation advocated by Hoover, 112

Council of National Defense, 34–35, 37, 63

Creel, George, 69, 229

Cummings, Homer, 223

Cummins, Albert B., 40

Daniels, Josephus, 18, 32, 41, 65–66, 74

Darrow, Clarence, 206

Davis, James J., 130

Davis-Kelly bill, 144

Dawson, Charles I., 188–89

Demobilization after World War I, 97

Department of Justice, 30, 33, 47, 129, 130, 132

Dickinson, John, 144–45

Doughton, Robert, 223

Economic structure of coal industry. *See* Structure of coal industry

Elections: 1932, 136–37; 1936, 228–30

Ellis, Wayne P., 200, 204, 207, 210

Energy: coal industry share, 122, 215

English Factory Act of 1833, 162

Excess capacity of coal industry, 21, 95, 112–13

Fair trade practices code advocated, 99

Federal Trade Commission, 27, 28, 29, 30, 39–40, 43, 49, 54, 84–85

Field, W. K., 60

Fireside chat of May 7, 1933, 145

Floods of 1927, 122

Ford, Henry, 181

Fort, John F., 42

Francis, James D., 60, 130, 141, 151–64, 185, 203

Freight rates, 14

Frick, H. C., Co., 154–55, 181–82

Fuel Administration, 56, 78, 82, 84–94, 91–92, 201–2

Fuelless Mondays, 65

Garfield, Harry A.: and industrial cabinet plan, 9, 98–101; appointed fuel administrator, 54–55; winter shutdown, 61, 62, 68–69; evaluated, 75, 80, 92–94; resigns, 91–92; on industrial cooperation, 98

Gary, Judge Elbert, 18

General Munitions Board, 36

Gifford, Walter S., 37, 39, 127–28

Gompers, Samuel, 28

Great Depression, 124–25, 139

Green, William, 104, 110, 175, 185, 227–28

Gregory, Thomas, 18

Grey, Benjamin, 152

Group-based legislation, 12. *See also* Guffey-Snyder and Guffey-Vinson acts

Guffey, Joseph F., 219, 291

Guffey-Snyder Act of 1935, 5, 141

Guffey-Vinson Act of 1937, 5, 219, 223

Hard, William, 72–75, 80

Harding, Warren, 114–15

Harlan County, Ky., 181

Harriman, Henry I., 143, 144
Hayden, Carl, 132
Hayes, Frank J., 98
Hearst, William R., 149
Henderson, Leon, 204, 209
Herrin massacre, 114
Hines, Walker, 143
Hitchcock, Gilbert, 66
Hoover, Herbert: on coal stabilization, 10, 112–13, 128, 133; recommends Garfield, 54; and Food Administration, 62; and trade associations, 109; and Great Depression, 126–28; on NIRA, 133; campaign of 1932, 136–37
Hosford, Charles F., 129, 131, 157, 233, 234
Howe, J. Wilson, 43
Hughes, Charles Evans, 221, 226–27
Hurley, Edward N., 18, 28–30
Hurley, Patrick, 187

Ickes, Harold, 148
Industrial cabinet plan of Garfield, 98–101
Industrial self-government, 9, 10, 11–12, 202–4
Inelastic demand and coal industry, 22
Injunctions and coal industry, 105, 187, 189
Internal Revenue Service, 90
Interstate Commerce Commission, 110

Jacksonville Agreement, 95, 119
Johnson, Hugh S.: on industrial self-government, 7, 9–11; and NIRA, 143–46; codifies industry, 148–50; and John L. Lewis, 152–53; on collective bargaining, 167; and captive mines, 176–77; on wage differentials, 186–87; on price fixing, 199–200; praises coal code, 204, 214; on Office Memorandum 228, 207; takes vacation from NRA, 208
Johnston, Forney, 157
Jones, Walter, 160

Kelly, Clyde, 130
Kennedy, John F., 245
Kenyon, William F., 111–12, 144
Kirstein, Louis E., 175
Knights of Labor, 26

Landon, Alfred, 230
Lane, Franklin K., 18, 27, 33, 38, 42–43, 103
Lamont, Robert, 128
Lauck, W. Jett, 123, 144
Lesher, C. E., 57–61
Lever, Asbury F., 45–46
Lever Act, 19, 48, 50, 53, 103
Lewis, David J., 132
Lewis, John L.: relations with Hugh Johnson, 52–53; statistician, 81; during 1920's, 95, 116, 118, 119, 123–24; on Davis-Kelly bill, 131; and NRA code, 136; and Franklin Roosevelt, 137, 175, 229; on NRA coal board, 173; at 1934 UMW convention 184–85; confronts Patrick Hurley, 188; attacks Clarence Darrow, 206; threatens operators over prices, 210–11; threatens strike over Guffey-Vinson bill, 222–24; seeks operator cooperation, 291
Lewis-Hayden proposal, 141
Lightless nights, 75
Lodge, Henry Cabot, 70
Lovett, Judge Robert S., 61, 62

McAdoo, William G., 65, 69, 70
McGrady, Edward, 155–56, 173
Manning, Van A., 57
Manufacturing shutdown, 1918, 65–66
Market areas: structure, 14, 15; established under code, 208; established under Guffey-Vinson Act, 234
Maynard Coal Co. v. *FTC*, 109
Merger movement, 25
Miner attitudes, 170–71
Mitchell, John, 37
Model code, 151
Moderwell, Charles M., 41
Moley, Raymond, 142–45

Morrow, James D. A., 60, 62, 71, 153–64, 185, 215, 217
Moses, Thomas, 154–56, 176
Munitions Standard Board, 36
Murray, Philip, 137, 168

National Association of Manufacturers, 145
National Bituminous Coal Commission: (first), 225; (second), 231, 234, 235, 238
National Bituminous Coal Industrial Board, 161, 199, 207, 211
National Bituminous Coal Labor Board, 172
National Bureau of Economic Research, 88
National Coal Association, 58–60, 80, 95, 110, 130, 140, 151–53, 212, 217, 291
National Coal Board of Arbitration, 212
National Industrial Recovery Act: prefigured, 112, 124, 129, 133–34; as law, 142–45; economic flaws in, 146; price-fixing provisions weak, 197
National Industrial Recovery Board, 209
National Labor Board, 174–75, 176
National Labor Relations Board, 177
National Recovery Administration: war origins, 10; organized, 147, 148, 149; economic impact on coal, 189–93, 195–96, 215–16, 236; price policy, 197–200, 213–16; price hearings, 205
National Recovery Review Board, 206
Nationalization of mines, 101
Neale, James B., 78–79
New Deal and analogue of war, 138, 139–40, 195, 201–2
Newlands, Francis J., 40, 51
Nims, Harry D., 56
North-South rivalry, 27, 97, 113, 116, 119, 120, 121–22, 140, 150–64, 218
Noyes, P. B., 65
Nye, Gerald, 205

O'Brian, John Lord, 129
Office Memorandum 228, 206–7
Office of Price Administration, 236
Ohio fuel crisis, 63
O'Neill, Charles, 140–41, 151, 185
Organized labor and Franklin Roosevelt, 229–30

Palmer, A. Mitchell, 103–7
Panic of 1907, 27
Peabody, Francis S., 18, 37–38, 45, 57, 70
Peale, Rembrandt, 59–60, 83
Penrose, Boise, 47
Perkins, Frances, 173
Pinchot, Gifford, 176
Planning Committee for Mineral Policy, 217
Pomerene, Atlee, 40, 46, 47
President's Reemployment Agreement, 148
Profit margins of coal: under Fuel Administration, 86–88; under NRA, 215
Progressivism, 240–42
Price fixing: alleged, 33; opposed, 49–50; under NRA, 213; under Guffey-Vinson Act, 232–36
Prices: stabilized under NRA, 12; slashed during prewar era, 25–26; during war, 32; under CCP, 39; under FA, 84–91; under code, 196

Quaker relief, 126, 128

Railroad Administration, 71
Rainey, Henry T., 30
Reading Formula, 175
Recession of 1921, 113
Red Scare, 101–7
Return of equity, 89–90
Return on investment, 108
Richberg, Donald, 11, 142–45, 156, 167, 177, 184, 207, 209, 214
Roche, Josephine, 153
Roosevelt, Franklin D.: on World War I experience, 7–8; and industrial self-government, 11; election of 1932, 136–39; and coal stabilization, 138, 143–46, 159,

161, 174, 177–78, 200, 205–6, 221, 223; and NIRA, 143–46; labor policies, 174–76, 177–78, 182; and NRA, 200, 205–6, 209, 221; and Guffey-Vinson Act, 221, 223; and organized labor in 1936 election, 228–30
Rummel, Judge H. D., 132
Rumsey, Mary, 200, 204

Sachs, Alexander, 148, 197
Sales agencies, 27, 28–29
"Save the Union" campaign of 1927, 121
Schechter Poultry Corp. v. *United States*, 189, 213, 221
Scott, Frank, 36
Self-government in industry. *See* Industrial self-government
Sherman Antitrust Act, 24–25
Shortages of coal, 63–64
Shutdown of manufacturing, 1918, 66–70
Simpson, Kenneth, 154, 156, 183
Smith, Blackwell, 209
Snead, Leonard A., 56, 61
Southern Appalachian area, 13
Southern Coal Association, 24
Southern operators, 187
Southern senators, 188
South-North rivalry. *See* North-South rivalry
Special Industrial Recovery Board, 147, 200
Specialized coals, 22
Steel industry and captive mines, 22, 154–56, 173–75
Steinmetz, Charles, 55
Strikes, 26, 101, 103–4, 110–12, 121–22, 154–56, 173–75, 187, 192–93, 222–24
Structure of coal industry: fragmented, highly competitive, complex, 5, 8, 13–14, 18, 21, 25, 27, 31, 34, 93–94, 169, 199, 200, 207, 210, 243–46; and war programs, 8; distinguished from anthracite coal, 18; and failure of New Deal programs, 8, 12, 232–33; and railroads, 21; impact of code on, 201. *See also* North-South rivalry;

United Mine Worker-northern operator alliance
Sunshine Anthracite Coal Co. v. *Adkins*, 233
Sutherland, George, 226
Swift and Company v. *United States*, 226
Swope, Gerard, 129, 175

Taggart, Ralph E., 185
Taplin, Frank E., 140, 169
Taylor, Henry N., 39, 59, 98
Taylor, Myron, 174
Taylor, Samuel A., 57
Teagle, Walter C., 175
Temporary National Economic Committee, 23, 90
Tidewater Coal Exchange, 38
Tillman, Benjamin, 23
Trade association movement, 24, 25, 129. *See also* Industrial self-government
Trade practices under code, 157–58, 160
Tugwell, Rexford, 142–45, 200
Tumulty, Joseph, 103
Turnblazer, William, 181
Tydings, Millard, 224

United Mine Worker-northern operator alliance, 5, 13, 14, 28, 92–93, 152–64, 183–93, 219–20, 228
United Mine Workers of America: collective bargaining, 26, 96–101, 103–4, 121–22, 144, 154–56, 173–75, 187, 192–93, 222–24; organizing strength, 81, 95, 168; stabilization bill of 1932, 130; and Franklin D. Roosevelt, 137; and North-South wage differential, 165; on Bituminous Coal Labor Boards, 180–82; and code authorities, 211; and 1936 election, 229; role in stabilization summarized, 243
United States Chamber of Commerce, 9, 28
United States Coal Commission, 90, 117–18
United States v. *E. C. Knight*, 226

Vardaman, James K., 46

Wages, 26, 125, 153, 156–57, 183, 190–93, 201. *See also* North-South rivalry; United Mine Workers of America, collective bargaining
Wagner, Robert F., 142–46, 175
Wallace, Henry A., 200
War Industries Board, 17, 72, 97–98
Warren, Bentley, 56
Warrum, Henry, 218, 233
Washington Agreement, 81–83, 96–101
Watson, James E., 123
Wheelwright, Jere H., 60

White, John P., 28, 83
Wiggin, Albert, 64
Williams, John Skelton, 74
Wilson, Edith Bolling, 106
Wilson, William B., 103–8
Wilson, Woodrow: and coal prices, 27, 44, 49, 86; relation with Harry Garfield, 55–56, 99–100; has stroke, 102, 106
Winter crisis, 1917–18, 63, 66–70
Wood, Fredrick H., 224

Young, Owen, 109

Zone plan of Fuel Administration, 72–74